SpringerBriefs in Molecular Science

SpringerBriefs in Molecular Science present concise summaries of cutting-edge research and practical applications across a wide spectrum of fields centered around chemistry. Featuring compact volumes of 50 to 125 pages, the series covers a range of content from professional to academic. Typical topics might include:

- A timely report of state-of-the-art analytical techniques
- A bridge between new research results, as published in journal articles, and a contextual literature review
- A snapshot of a hot or emerging topic
- An in-depth case study
- A presentation of core concepts that students must understand in order to make independent contributions

Briefs allow authors to present their ideas and readers to absorb them with minimal time investment. Briefs will be published as part of Springer's eBook collection, with millions of users worldwide. In addition, Briefs will be available for individual print and electronic purchase. Briefs are characterized by fast, global electronic dissemination, standard publishing contracts, easy-to-use manuscript preparation and formatting guidelines, and expedited production schedules. Both solicited and unsolicited manuscripts are considered for publication in this series.

More information about this series at http://www.springer.com/series/8898

Tatsuhisa Kato · Naoki Haruta · Tohru Sato

Vibronic Coupling Density

Understanding Molecular Deformation

Springer

Tatsuhisa Kato
Fukui Institute for Fundamental Chemistry
Kyoto University
Sakyo-ku, Kyoto, Japan

Naoki Haruta
Fukui Institute for Fundamental Chemistry
Kyoto University
Sakyo-ku, Kyoto, Japan

Tohru Sato
Fukui Institute for Fundamental Chemistry
Kyoto University
Sakyo-ku, Kyoto, Japan

ISSN 2191-5407 ISSN 2191-5415 (electronic)
SpringerBriefs in Molecular Science
ISBN 978-981-16-1795-9 ISBN 978-981-16-1796-6 (eBook)
https://doi.org/10.1007/978-981-16-1796-6

This Springer imprint is published by the registered company Springer Nature Singapore Pte Ltd.
The registered company address is: 152 Beach Road, #21-01/04 Gateway East, Singapore 189721, Singapore

Preface

Nobel laureate Roald Hoffmann pointed out, "Understanding and Explaining and their Strong-Tie to Teaching" in the memorial lecture at the Kenichi Fukui 100th birthday anniversary 2018 in Kyoto.[1] He sketched a possible path to coexistence, way from theory and understanding through simulation and then back again to understanding. We want to quote his sketches. He gave the process of understanding, insight, and explanation, and then prediction. Understanding is often tacit, or silent, a state of mind. It is usually qualitative, though it may have quantitative aspects. Quantitative reinforces qualitative. And simulation is around the corner. An explanation is inherently more pedagogic and storytelling. More useful to science are hypotheses, alternative narratives. An explanation comes to us through stories, chemical stories. Prediction is the conceptual passage between understanding and simulation and is the practical counterpart of contemplative understanding, "I know how" instead of "I know why."

A molecule can be regarded as a system of electrons in a framework of nuclei, that is, a molecular structure. Chemists are interested in molecular motions under the interactions of other molecules or an electromagnetic field. The motion of a molecule is decomposed into translational, rotational, and vibrational modes. An intramolecular motion, i.e., molecular deformation, is expressed as a combination of vibrational modes.

Exactly speaking, vibrations and motions of electrons cannot be separated. Therefore, any change of an electronic state gives rise to a molecular deformation. The couplings among vibrations and electrons are called vibronic coupling (VC). The magnitude of deformation depends on the strength of VCs, vibronic coupling constants (VCCs). The direction of deformation depends on the relative values of the VCCs of vibrational modes. Thus, if we can understand the reason for the relative strength and magnitude of VCCs, we can explain the molecular deformation under a certain interaction. Vibronic coupling density (VCD), a density form of a VCC,

[1]Roald Hoffmann, "Simulation versus Understanding: A tension, and not just in Quantum Chemistry.", the lecture in the Memorial Symposium of "Kenichi Fukui 100th birthday anniversary," Kyoto, 2018.

enables us to explain such a reason. In this monograph, we will give the instructive path to the VCD and the VCC analyses.

Chemistry in alchemy was just aimed at getting gold without understanding. Modern chemistry emerged with the concept of atoms and molecules. After the quantum study, the electron behavior circulating nuclei was led to the principal concept underlying all explanations in chemistry. Many textbooks have given the plausible explanations to clarify the molecular structure. And the frontier molecular orbital concepts were proposed to visualize the path of a chemical reaction. The conventional explanations have provided students with a considerable familiarity with the molecular structure in terms of the electronic state. However, the more rational and more convincing ways should be given. Here the VCD and the VCC analyses are introduced. They are starting from the *ab initio* molecular Hamiltonian, and systematic, rational ways to understand chemical phenomena, and which can give the quantitative evaluation of the force applied under the chemical deformation process. We offer the guidelines to integrate the traditional "hand-waving" approach of chemistry with more rational and general VCD and VCC alternative. Further outlooks for the newly functionalized chemical systems. Thus, through the visualization by VCD and the evaluation by VCC, the study of chemistry by molecular orbital theory is brought into the domain of substantial science, where qualitative concepts can be rendered quantitatively and tested rigorously against the quantum theory.

Kyoto, Japan
February 2021

Tatsuhisa Kato
Naoki Haruta
Tohru Sato

Acknowledgements

The authors are greatly indebted to students who made this book possible: Ken Tokunaga, Motoyuki Uejima, Katsuyuki Shizu, Naoya Iwahara, and Yuichiro Kameoka. TS acknowledges colleagues for guiding his theoretical studies: Kazuyoshi Tanaka, Arnout Ceulemans, and Liviu F. Chibotaru, and for discussions on interpretations of experimental results: Hironori Kaji, Tsunehiro Tanaka, Kentaro Teramura, Saburo Hosokawa, Kenji Matsuda, and Takashi Hirose.

Contents

Acronyms

AO	Atomic Orbital
AVCC	Atomic Vibronic Coupling Constant
CASSCF	Complete Active Space Self-consistent Field
DFT	Density Functional Theory
EVCD	Effective Vibronic Coupling Density
HOMO	Highest Occupied Molecular Orbital
irrep	irreducible representation
JT	Jahn–Teller
JTE	Jahn–Teller Effect
LUMO	Lowest Unoccupied Molecular Orbital
MO	Molecular Orbital
OVCC	Orbital Vibronic Coupling Constant
OVCD	Orbital Vibronic Coupling Density
PJTE	Pseudo-Jahn–Teller Effect
QVCC	Quadratic Vibronic Coupling Constant
QVCD	Quadratic Vibronic Coupling Density
SOMO	Singly Occupied Molecular Orbital
TDMD	Transition Dipole Moment Density
VB	Valence Bond
VC	Vibronic Coupling
VCC	Vibronic Coupling Constant
VCD	Vibronic Coupling Density
VSEPR	Valence Shell Electron Pair Repulsion

Symbols

$\boldsymbol{x}$	Cartesian coordinate in the three-dimensional space
x, y, z	x, y, z-components of $\boldsymbol{x}$, relative coordinates (Chap. 2.2)
X, Y, Z	Coordinates of the center of mass (Chap. 2.2)
$\boldsymbol{r}$	A set of spatial coordinates of all electrons
i, j	Labels of electrons, Labels of atom (Chap. 2.2)
l_i	Bond length of atom i and $i+1$
φ_i	Bond angle of atom $i-1$, i, and $i+1$
χ_i	Dihedral angle of atom $i-1$, i, $i+1$, and $i+2$
$\boldsymbol{r}_i$	Spatial coordinate of electron i
x_i, y_i, z_i	x, y, z-components of $\boldsymbol{r}_i$
$\boldsymbol{\omega}$	A set of spin coordinates of all electrons
ω_i	Spin coordinate of electron i
$\boldsymbol{R}$	A set of spatial coordinates of all nuclei
A, B	Labels of nuclei
$\boldsymbol{R}_A$	Spatial coordinate of nucleus A
X_A, Y_A, Z_A	x, y, z-components of $\boldsymbol{R}_A$
$\boldsymbol{R}^0$	Reference nuclear configuration
$\boldsymbol{R}_A^0$	Spatial coordinate of nucleus A in the reference nuclear configuration
X_A^0, Y_A^0, Z_A^0	x, y, z-components of $\boldsymbol{R}_A^0$
$\Delta\boldsymbol{R}$	Displacement of all nuclei from $\boldsymbol{R}^0$
$\Delta\boldsymbol{R}_A$	Displacement of nucleus A from $\boldsymbol{R}_A^0$
$\Delta X_A, \Delta Y_A, \Delta Z_A$	x, y, z-components of $\Delta\boldsymbol{R}_A$
$\boldsymbol{Q}$	A set of all mass-weighted vibrational coordinates
α, β	Labels of vibrational modes
Q_α	Mass-weighted vibrational coordinate of mode α
q_α	Vibrational coordinate of mode α in the real space
$\boldsymbol{u}_\alpha$	Vibrational vector of mode α in the mass-weighted space
$\boldsymbol{v}_\alpha$	Vibrational vector of mode α in the real space
μ_α	Reduced mass of mode α
ω_α	Frequency of mode α
$v_\alpha(\boldsymbol{x})$	Linear potential derivative with respect to vibrational mode α

$w_{\alpha\beta}(\boldsymbol{x})$	Quadratic potential derivative with respect to vibrational modes α, β
$w_{\alpha_1\cdots\alpha_k}(\boldsymbol{x})$	k-th-order potential derivative with respect to vibrational modes $\alpha_1 \cdots \alpha_k$
m, n	Labels of electronic states
$\Psi_m(\boldsymbol{r}, \boldsymbol{R})$	m-th total electronic wave function at the nuclear configuration $\boldsymbol{R}$
$E_m(\boldsymbol{r}, \boldsymbol{R})$	m-th total electronic energy at the nuclear configuration $\boldsymbol{R}$
$\rho_m(\boldsymbol{x})$	Electron density in the m-th electronic state
$\Delta\rho_m(\boldsymbol{x})$	Electron density difference between the reference and the m-th electronic states
$\rho_{mn}(\boldsymbol{x})$	Overlap density between m-th and n-th electronic states
$\rho_{ab}(\boldsymbol{x})$	Overlap density between a-th and b-th molecular orbitals
$V_{mn,\alpha}$	Linear vibronic coupling constant between m-th and n-th electronic states with respect to vibrational mode α
$W_{mn,\alpha\beta}$	Quadratic vibronic coupling constant between m-th and n-th electronic states with respect to vibrational modes α, β
$W_{mn,\alpha_1\cdots\alpha_k}$	k-th-order vibronic coupling constant between m-th and n-th electronic states with respect to vibrational modes $\alpha_1 \cdots \alpha_k$
$\eta_{mn,\alpha}(\boldsymbol{x})$	Linear vibronic coupling density between m-th and n-th electronic states with respect to vibrational mode α
$\eta_{mn,\alpha\beta}(\boldsymbol{x})$	Quadratic vibronic coupling density between m-th and n-th electronic states with respect to vibrational modes α, β
$\eta_{mn,\alpha_1\cdots\alpha_k}(\boldsymbol{x})$	k-th-order vibronic coupling density between m-th and n-th electronic states with respect to vibrational modes $\alpha_1 \cdots \alpha_k$
$V_{ab,\alpha}$	Linear orbital vibronic coupling constant between a-th and b-th molecular orbitals with respect to vibrational mode α
$W_{ab,\alpha\beta}$	Quadratic orbital vibronic coupling constant between a-th and b-th molecular orbitals with respect to vibrational modes α, β
$W_{ab,\alpha_1\cdots\alpha_k}$	k-th-order orbital vibronic coupling constant between a-th and b-th molecular orbitals with respect to vibrational modes $\alpha_1 \cdots \alpha_k$
$\eta_{ab,\alpha}(\boldsymbol{x})$	Linear orbital vibronic coupling density between a-th and b-th molecular orbitals with respect to vibrational mode α
$\eta_{ab,\alpha\beta}(\boldsymbol{x})$	Quadratic orbital vibronic coupling density between a-th and b-th molecular orbitals with respect to vibrational modes α, β
$\eta_{ab,\alpha_1\cdots\alpha_k}(\boldsymbol{x})$	k-th-order orbital vibronic coupling density between a-th and b-th molecular orbitals with respect to vibrational modes $\alpha_1 \cdots \alpha_k$

Chapter 1
Qualitative Explanation of Molecular Structures by Various Approaches

Abstract Valence bond (VB) picture and valence shell electron pair repulsion principle (VSEPR) [1–3] are localized bond approaches. VB picture is the historic landmark for modern chemistry, which is based on the valence bond. It invests each atom with the valence according to the number of electron, which is the basic concept for chemical bonds of the molecule; however, it has the fatal fault *in which hybridization of the bonding orbital is determined after empirically knowing the molecular structure*. VSEPR is more applicable to elucidate various molecular structures but is still an empirical picture. Both have no proper way to explain molecular structure in the excited state. On the other hand, molecular orbital (MO) theory is the approach based on delocalized electrons. It starts from the molecular Hamiltonian, and is a more systematic and rational way to understand molecular structures. Many textbooks have given explanations for the molecular deformation [4]. In this chapter, several well-known issues are discussed, the bond elongation of ethylene under the anionization, the bent structure of water, the nonplanar structure of NH_3, the tautomerism between benzenoid and quinoid forms of organic molecule, the triangle structure of C_3H_3, and the chemical process of Diels–Alder reactions.

Keywords Valence bond · Valence shell electron pair repulsion principle · Molecular orbital · Anionization · Walsh diagram · Pseudo-Jahn-Teller theory · Renner-Teller theory · Highest occupied molecular orbital · Lowest unoccupied molecular orbital · Tautomerism · Benzenoid · Quinoid · Jahn-Teller effect · Frontier orbital theory · Diels-Alder reaction

1.1 Elongation of Bond on Anionization of H_2 and C_2H_4

Here molecular orbitals of hydrogen are shown in Fig. 1.1. It is well known that when a neutral hydrogen molecule acquires an electron, and the chemical bond is elongated, since the additional electron occupies the anti-bonding $1\sigma_u$ lowest unoccupied molecular orbital (LUMO).

T. Kato et al., *Vibronic Coupling Density*,
SpringerBriefs in Molecular Science,
https://doi.org/10.1007/978-981-16-1796-6_1

Fig. 1.1 Molecular orbitals of hydrogen molecule

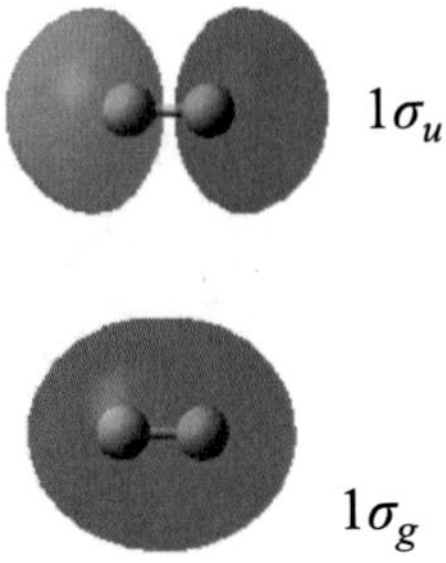

Anionization of ethylene also gives rise to an elongation of the double bond in the process of the electron occupation of the anti-bonding π-orbital. The bonding MO is composed of the in-phase mixing of atomic orbitals, which gives the positive overlap of the electronic wave function resulting in an attractive force between nuclei. On the other hand, the anti-bonding MO exhibits a repulsive force between nuclei. In the case of hydrogen, the effect is partly because an electron occupying the anti-bonding orbital is excluded from the internuclear region, and hence is distributed largely outside the bonding region. In effect, whereas an electron occupying the bonding orbital pulls two nuclei together, an electron occupying the anti-bonding orbital pulls the nuclei a part [5]. However, in the case of ethylene, no substantial interpretation of the repulsive force is given, nor the quantitative estimation of the force so far in both cases. The vibronic coupling density (VCD) analysis will give the exact physical picture of the repulsive force in Sect. 3.1.

1.2 Bent Deformation of H_2O and NH_2

The VSEPR approach [1–3] gives an interpretation for the bent molecular structure of H_2O as the following. An oxygen atom has six electrons occupying its outermost shell. Each of the two H atoms donates one electron to the O atom, locating eight electrons around the O atom. Thus, four pairs of the electrons would be distributed in a tetrahedral fashion around the central atom if the potential around the O atom was isotropic. Among them, two pairs form two O–H bonds while the other two pairs remain unused, which are called lone pairs. Due to repulsion among the bonding electrons and lone pairs, the bond angle H–O–H is not exactly tetrahedral (109° 28′). Thus, H_2O becomes a bending molecule with the H–O–H angle of 104°. However, VSEPR approach just gives the interpretation of the actual H–O–H angle of 104° less than 109° 28′ in tetrahedral under the empirical assumption of H_2O molecule's electronic structure as shown in Fig. 1.2, not the interpretation why the molecule becomes bent.

More reasonable interpretation for the bent structure of H_2O has been proposed by using the *Walsh diagram* [6] based on the MOs in Fig. 1.2. The *Walsh diagram* is the correlation diagram classifying the molecular orbitals of H_2A according to their

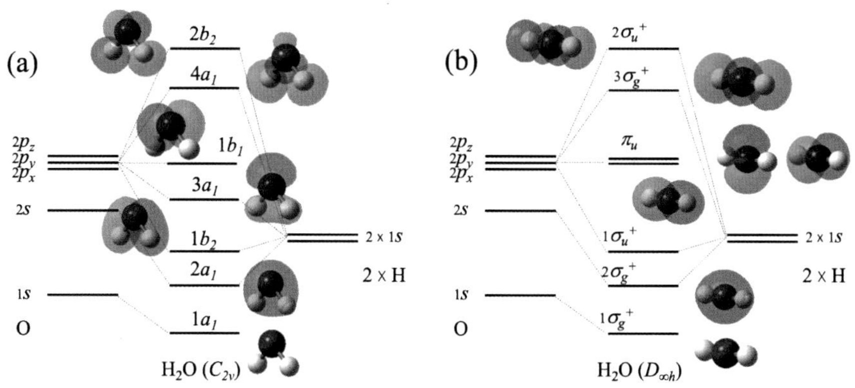

Fig. 1.2 **a** Molecular orbitals of H_2A molecule in bent structure of C_{2v} symmetry. **b** Molecular orbitals of H_2A molecule in linear structure of $D_{\infty h}$ symmetry

symmetries, which leads to the primitive approach to predict the structure for H_2A molecule, and suggests the useful approach with the orbital symmetry. Supposing the linear structure of H_2A, the HAH angle is 180°, and the point group of the molecule is $D_{\infty h}$. The MOs in the linear structure are shown in Fig. 1.2b. The fourth and fifth orbitals are two degenerate π_u-orbitals composed of p_z and p_y atomic orbitals of the central oxygen. When the H–A–H angle is decreasing from 180° to 90°, the former π_u (p_z)-orbital interacts with the upper unoccupied $3\sigma_g^+$-orbital and has an overlap with the symmetric pair of hydrogen orbitals. As a result, the π_u (p_z)-orbital is stabilized with the decrease of the H–A–H angle as shown in Fig. 1.3. The other orbital of π_u (p_y) remains at the same energy level. On the other hand, the orbital energies of lower lying σ_u and σ_g increase with the decrease of the overlap of p_x with hydrogen orbitals. As a result, the correlation diagram can be drawn according to their symmetries as shown in Fig. 1.3, which applies to the molecule of type H_2A. When the diagram is applied to H_2O, and eight valence electrons are configured on MOs, all four MOs are occupied, and the configuration would be stabilized at the angle of 104.5°, not at 90° nor 180°. In the case of NH_2, seven electrons occupy the four orbitals, doubly occupy lower-lying three orbitals and singly occupy the fourth orbital. Then the electronic state of NH_2 is also stabilized at the deformed bent structure by the same mechanism.

Walsh diagram is intuitive and points out that the symmetry of MO plays a decisive part of determining molecular structures, but is quite qualitative. Going back to the molecular Hamiltonian to know the origin of molecular deformation, the other approach of the pseudo-Jahn–Teller theory [7] and the Renner–Teller theory [8] gives more strict discussion on the symmetry of MO with respect to the deformation coordinate. The mechanism that the ground state of H_2O is stabilized with the deformation coordinate will be explained by the pseudo-Jahn–Teller effect in Sect. 3.2. In the case of the linear molecule NH_2, the degenerate states are spontaneously split with respect to the out of axis deformation by the Renner–Teller effect, also explained in Sect. 3.2.

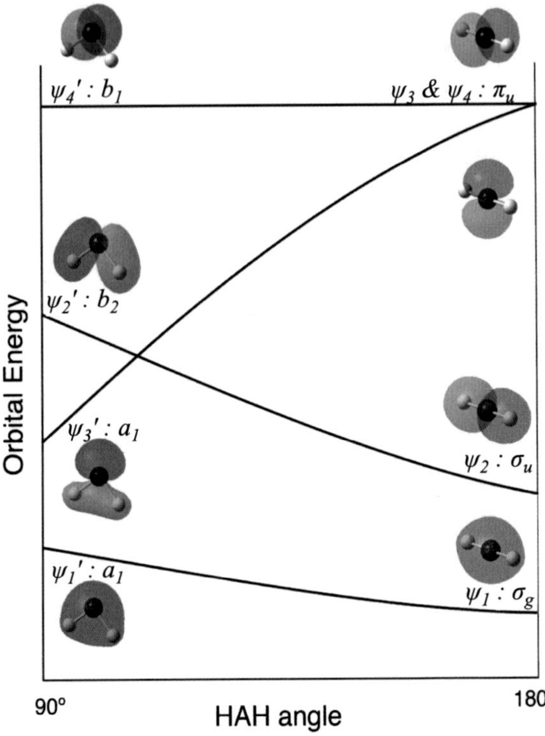

Fig. 1.3 Walsh diagram: the correlation diagram of MO energies according to their symmetries

1.3 Nonplanar Molecular Structure of NH_3

The nonplanar structure of NH_3 has been understood in the VSEPR scheme [1–3]. Nitrogen atom has five electrons which occupy its outermost shell. Each of the three H atoms donate one electron to the N atom, locating eight electrons around the N atom. Thus, the four pairs of electrons would be distributed in a tetrahedral fashion around the central atom. Three pairs form the three N–H covalent bonds, while the fourth pair remains unused as a bond. The fourth pair is called lone pair. Due to the difference of repulsion between the pair of electrons consisting of covalent bond and the lone pair, the bond angle H–N–H is not exactly tetrahedral (109° 28′) but is 107.8°. Thus, NH_3 is understood as a trigonal pyramidal molecule. However, this scheme of understanding is again based on the assumption that three N–H bonds are not aligned in-plane, and just describes the reason why the angle H–N–H is not an exact tetrahedral angle. There is no explanation of the force to make the structure bent.

Here, we can explained for the bent structure of NH_3 by using Walsh diagram as shown in Fig. 1.5. Assuming the planar structure of NH_3, the point group of the molecule is D_{3h}. The MO of the valence electrons is shown in Fig. 1.4. The orbital of a_2'' is the HOMO (highest occupied molecular orbital) of NH_3 and the orbital $2a_1'$ is

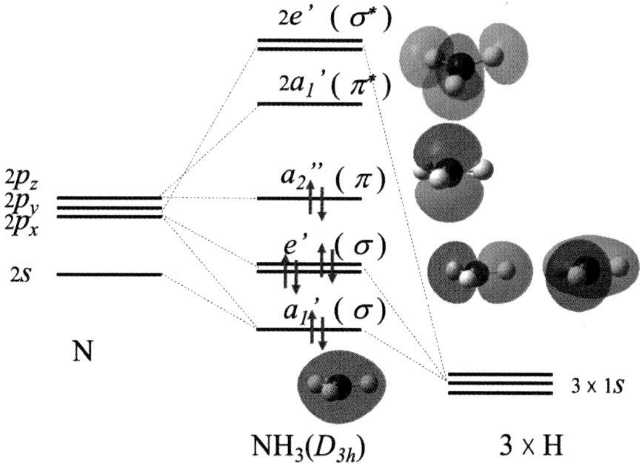

Fig. 1.4 Molecular orbitals of NH_3 with the planar structure of D_{3h}

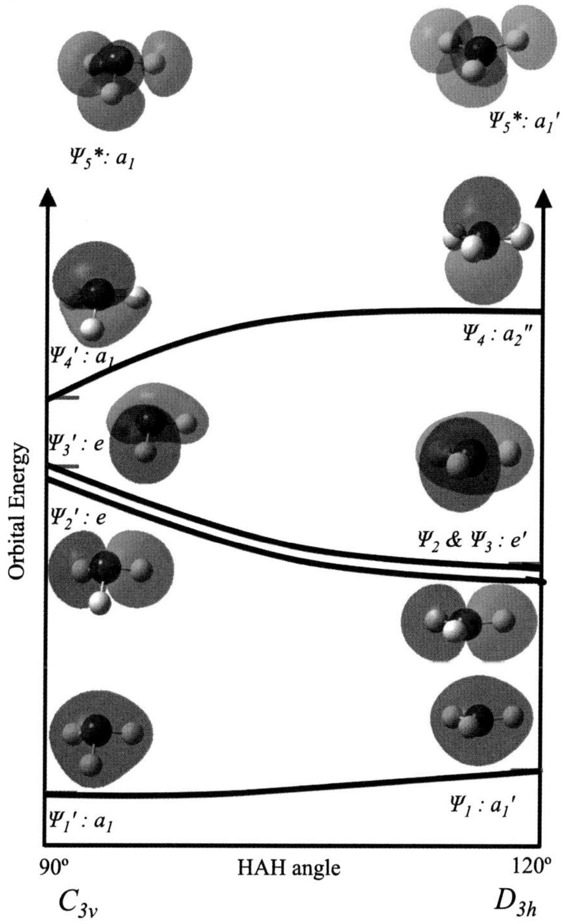

Fig. 1.5 Walsh diagram of NH_3

the LUMO. Drawing the correlation of MO energies according to their symmetries, the diagram shown in Fig. 1.5 is given with a planar and pyramidal configuration. The HOMO orbital is well stabilized with changing from the a_2''-orbital in the planar structure to the a_1 in the pyramidal structure because of the resonance interaction of p_z-orbital with the LUMO $2a_1'$. On the other hand, the energy levels of degenerate e-orbitals are increasing since the overlap of p_x- or p_y-orbital with the corresponding combination of the hydrogen decreases. When eight valence electrons occupy the four MOs, the minimum sum of orbital energies is fulfilled at some position of the pyramidal structure. The Walsh diagram for NH_3 is again entirely qualitative, and more systematic interpretation will be given by the pseudo-Jahn–Teller theory [7] in Sect. 3.3.

1.4 Zigzag Conformation of Cycloparaphenylene

The cycloparaphenylene (CPP) is the cyclic molecule that consists of poly-phenylene connecting in the para positions. We can regard it as the shortest carbon nanotube. For example, the shortest unit of the (6, 6) armchair carbon nanotube is 6-CPP, as shown in Fig. 1.6. Then, it is usually supposed to be a belt-like molecule. However, it is not the case. Experimental data and theoretical calculation show that the most stable molecular structure is not a belt-like structure but an alternating zigzag conformation in Fig. 1.7. The X-ray diffraction of 6-CPP reported a non-zero dihedral angle C_{ortho}–C_{ipso}–C_{ipso}–C_{ortho} [9], which is defined in Fig. 1.7, and the MO calculation predicted the dihedral angle of 15° [10].

The interconversion between belt-like and alternating zigzag conformations is described in organic-chemical point of view as the tautomerism between benzenoid and quinoid forms as shown in Fig. 1.8. This is the intuitive way of the explanation

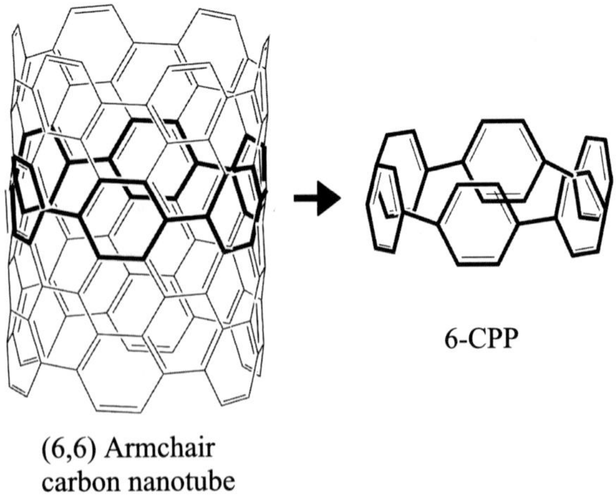

Fig. 1.6 6-CPP as the shortest unit of the (6, 6) armchair carbon nanotube

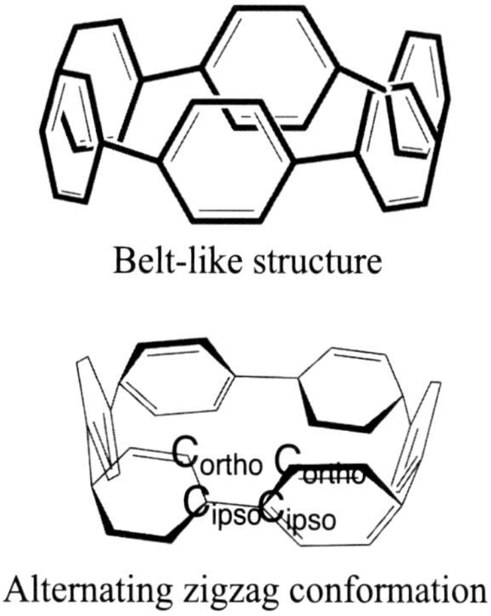

Fig. 1.7 Upper: belt-like structure of 6-CPP, lower: alternating zigzag conformation. C_{ortho} and C_{ipso} carbons are those at the designated positions of molecule, the angle formed by C_{ortho}, C_{ipso}, C_{ipso}, and C_{ortho} is defined as a dihedral angle C_{ortho}–C_{ipso}–C_{ipso}–C_{ortho}

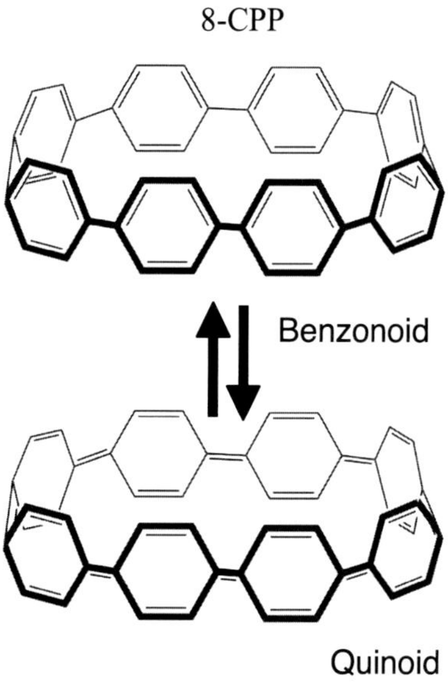

Fig. 1.8 Tautomerism between benzenoid and quinoid forms

of the electron delocalization. The X-ray diffraction of 8-CPP reported that almost equivalent bond lengths in benzene unit [11], and concluded that the approximately equal bond lengths indicate that a benzenoid structure is preserved in 8-CPP. As a result, the bond between neighboring benzene units has a single bond nature and the dihedral angle is not necessarily zero. On the other hand, the dihedral angle of the quinoidal form is forced to be zero because of the double bond nature between neighboring benzene units. However, the description of tautomerism just classifies the fashion of the electron delocalization. The strain energy with the bend angle at

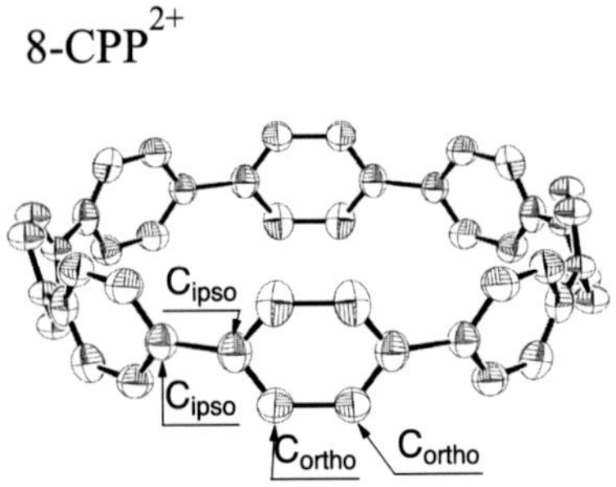

Fig. 1.9 ORTEP figure of the single crystal of 8-CPP^{2+} [13]

the C_{ipso} carbon and the steric interactions between neighboring C_{ortho} carbons have been usually described for the force leading to each form [12].

The 8-CPP radical cation and dication have recently been isolated and characterized [13] as shown in Fig. 1.9. The results of single-crystal X-ray analysis showed a significant decrease in the averaged dihedral angle between the neighboring para phenylene units of the dication species as compared to that observed in the neutral species [11]. The analysis also revealed that the C_{ipso}–C_{ipso} and C_{ortho}–C_{ortho} bonds shorten, while the C_{ipso}–C_{ortho} bonds of the dication elongate as compared to those in the neutral species. In other words, this signifies a change in the bond-alternation pattern into the quinoid form. This interconversion from the benzenoid to the quinoid form due to the ionization would indicate that the zigzag conformation is originated from the electronic structure of the molecule, not from the steric hindrance between neighboring C_{ortho} carbons. In Sect. 3.4, the explanation by using the VCD is given for the origin of the zigzag conformation of 6-CPP.

1.5 Stable Structure of C_3H_3

Jahn and Teller found that all the non-linear nuclear configurations are unstable for an orbitally degenerate electronic state [14]. This effect was named after Jahn and Teller, called Jahn–Teller effect. The Jahn–Teller effect is usually referred as the mechanism explaining the ligand deformation of a metal complex in inorganic chemistry.

For example, a d^9 metal ion of Cu^{2+} is octahedrally coordinated, the metal d_{xy}-, d_{yz}-, and d_{xz}-orbitals are equivalent and involve electron density between the axes containing both the metal ion and the ligands. Both of the d_z^2- and $d_x^2 - y^2$-orbitals direct electron density toward the ligands. The octahedral arrangement of ligands around the metal ion splits the five d-orbitals into two sets, as shown in Fig. 1.10, one set (t_{2g}) being triply degenerate and the other (e_g) being doubly degenerate. The t_{2g}-orbitals are stabilized, and the e_g-orbitals are destabilized relative to their energies in a spherical field, the energy difference between the t_{2g}- and e_g-orbitals being designated Δ_0 as shown in Fig. 1.10. The driving force of the Jahn–Teller distortion

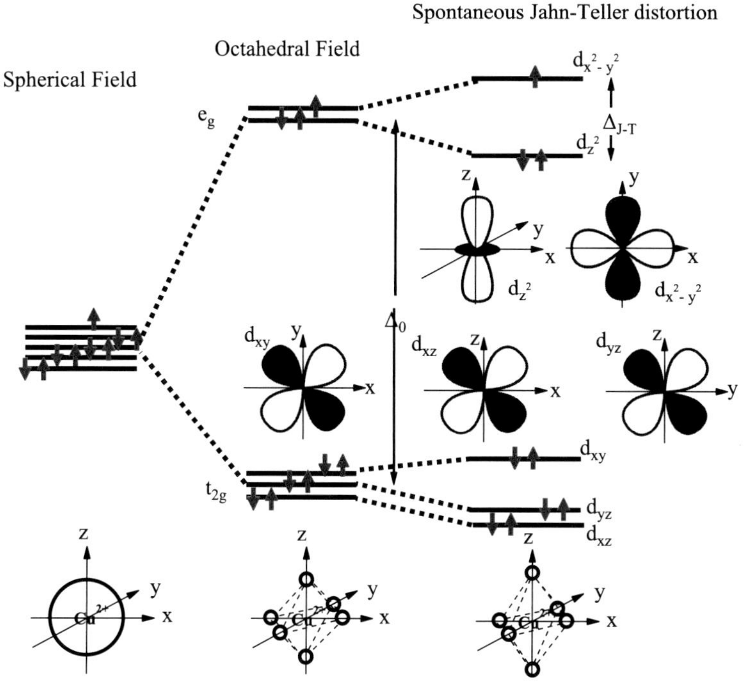

Fig. 1.10 The electronic energy levels for Cu^{2+} in a spherical field (left), an octahedral field (middle), and an elongated octahedral field (right)

of Cu^{2+} in octahedral ligand results from the unequal occupancies of the two e_g-orbitals, which are split by the distortion shown in Fig. 1.10. Where the Jahn–Teller distortion occurs, the singly occupied orbital is destabilized by the same energy as the doubly occupied orbital is stabilized, resulting in a split by the energy of Δ_{JT}. Ligand-field arguments show that single occupancy of the d_z^2- and $d_x^2 - y^2$-orbitals of Cu^{2+} will result in compressed and elongated octahedral ligand, respectively.

The structure of an organic molecule, for example, C_3H_3 can also be predicted from the Jahn–Teller effect (JTE), which explains the deformation from the equilateral triangle structure of C_3H_3 [15]. Let us consider the electronic state of C_3H_3 in the simple Hückel MO theory using valence π-orbitals. In the equilateral triangle structure of the D_{3h} symmetry, three π-orbitals of e''_θ, e''_ϵ, and a''_1 are composed as shown in Fig. 1.11. The e''_θ- or e''_ϵ-orbital is partially occupied in the ground electronic configuration resulting in the doubly degenerated ground electronic state. Among 12 vibrational normal modes, shown in Fig. 3.12, the E' modes give rise to the deformation causing the energy split for the doubly degenerated ground electronic state. The E' modes are called Jahn–Teller active modes as referred in Sect. 3.5. Figure 1.11 shows the deformation along the $E'(2)$ mode which stabilizes the orbital energy of the e''_θ-orbital, which is symmetric with one plane σ_v perpendicular to the molecular plane, and destabilizes that of the e''_ϵ-orbital, anti-symmetric with σ_v. And vice versa

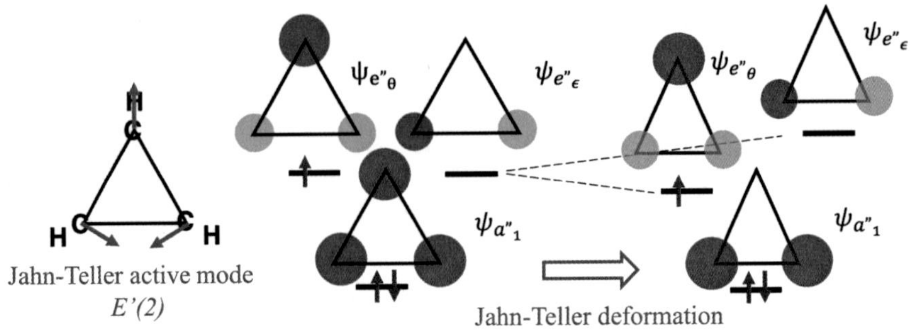

Fig. 1.11 The Jahn–Teller effect of C_3H_3

in the opposite direction of the $E'(2)$ vibration. This is the simple interpretation that the $E'(2)$ deformation causes the energy split for the doubly degenerated ground electronic state.

1.6 Frontier Orbital Theory for Diels–Alder Reaction

One of the central issues in chemistry is regioselectivity: which region is reactive in a molecule. The scheme of the HOMO–LUMO interaction between reactants is stated by the fundamental proposition that a majority of chemical reactions should take place at the position and in the direction of maximum overlapping of the HOMO and the LUMO of the reactants [16].

For example, the preferable path of the Diels–Alder reaction between ethylene and butadiene is explained as the approach between the HOMO of butadiene and the LUMO of ethylene giving the maximum overlapping shown in Fig. 1.12.

Since molecular deformations play an essential role in chemical reactions, the concept of VCD can also apply for the regioselectivity problems in chemical reactions.

In the VCD picture of the Diels–Alder reaction, we can apply the similar isolated reactant approach for larger and complicated reaction systems as of styrene and of C_{60} in Sect. 3.6. We will define the effective vibronic coupling density (EVCD) to clarify the regioselectivity of the Diels–Alder reaction with ethylene to C_{60}.

Fig. 1.12 Diels–Alder reaction

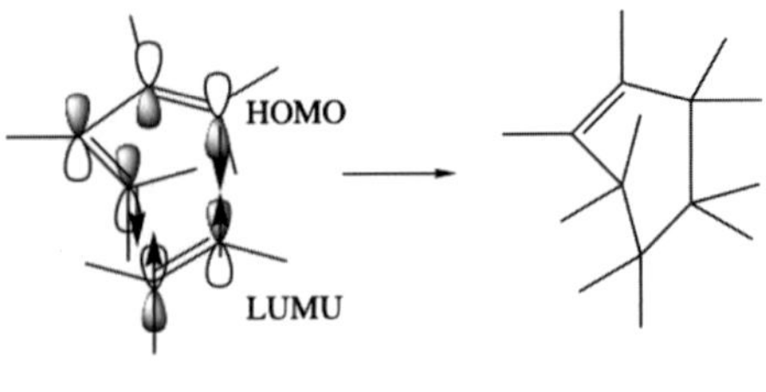

References

1. Gillespie RJ, Nyholm RS (1957) Quart Rev Chem Soc 11:339
2. Gillespie RJ (1963) J Chem Educ 40:295
3. Gillespie RJ (2008) Coord Chem Rev 252:1315
4. Schastnev PV, Schegoleva LN (1995) Molecular distortions in ionic and excited states. CRC Press Inc, New York
5. Atkins P, de Paula J (2006) Physical chemistry, 8th edn. W. H. Freeman and Company, New York
6. Walsh AD (1953) J Chem Soc 2260–2331
7. Bersuker I, Gorinchoi N, Polinger V (1984) Theo Chim Acta 66:161
8. Renner R (1934) Z Physik 92:172
9. Xia J, Jasti R (2012) Angew Chem Int Ed 51:2474
10. Wong BM (2009) J Chem Phys C 113:21921
11. Xia J, Bacon JW, Jasti R (2012) Chem Sci 3:3018
12. Bachrach SM, Stück D (2010) J Org Chem 75(19):6595
13. Kayahara E, Kouyama T, Kato T, Takaya H, Yasuda N, Yamago S (2013) Angew Chem Int Ed 52:13722
14. Jahn HA, Teller E (1937) Proc R Soc Lond A 161:220
15. Kayi H, Garcia-Fernandez P, Bersuker IB, Boggs JE (2013) J Phys Chem A 117:8671
16. Fukui K (1971) Acc Chem Res 4(57)

Chapter 2
Molecular Deformations and Vibronic Couplings

Abstract We define a molecular deformation, and discuss how to specify molecular deformations in a diatomic and triatomic molecules A_3 and A_2B taken as examples. An approximate molecular Hamiltonian for a polyatomic molecule is introduced. The origins of molecular deformations, vibronic couplings, are discussed for non-degenerate electronic states and degenerate electronic states (Jahn–Teller effect, pseudo-Jahn–Teller effect, and Renner–Teller effect).

Keywords Jahn–Teller effect · Pseudo-Jahn–Teller effect · Renner–Teller effect · Herzberg–Teller expansion · Crude adiabatic representation · Reduced mass · Moment of inertia · Angular momentum · Centrifugal deviation · Translational mode · Rotational mode · Vibrational mode · Euler angle · Group theory · Generator · Representation · Character · Irreducible representation · Cartesian coordinate · Symmetry-adapted coordinate · Bond angle · Dihedral angle · Z-matrix · Normal mode · Vibrational mode · Mass-weighted coordinate · Vibronic Hamiltonian · Linear vibronic coupling constant · Quadratic vibronic coupling constant · Diabatic representation · Adiabatic representation · Wigner–Eckart theorem · Clebsch–Gordan coefficient

2.1 Introduction: Molecular Deformations and Its Driving Force

In the previous chapter, we have glanced at some conventional explanations on the origins of molecular deformations. This chapter devotes defining molecular deformations and its driving force, vibronic coupling. Furthermore, their driving forces are analyzed in terms of VCD concepts in the succeeding chapter.

In this book, we define a molecular deformation as a change of the internal coordinates of a molecule from a certain reference geometry. We assume a motion of a molecule as separated modes: translation, rotation, and vibration. Within the assumption, an internal coordinate can be identified as a vibrational coordinate. The separation is illustrated in Sect. 2.2 taking a diatomic molecule as an example.

T. Kato et al., *Vibronic Coupling Density*,
SpringerBriefs in Molecular Science,
https://doi.org/10.1007/978-981-16-1796-6_2

When we discuss a molecular deformation, we assume some starting structure. We call the structure a reference geometry. Such a structure may be an equilibrium structure for an initial electronic state Ψ_0 or a hypothetical structure, the symmetry of which is sometimes higher than the real one. A molecular deformation upon a change of the electronic state is a real process and could be observable in the former case. For example, the bond elongations of H_2 and C_2H_4 discussed in Sect. 1.1 are a real process. However, the instability of NH_3 discussed in Sect. 1.3 is not a real but hypothetical process just for the explanation. The starting electronic state Ψ_0 is called a reference state. Note that the concept of the reference state includes its structure as reference geometry.

In both cases, the driving forces can attribute to the vibronic coupling. A concept of the vibronic coupling can be understood starting from the Herzberg–Teller expansion and crude adiabatic representation. Sections 2.6 and 2.7 devote to the derivation of the vibronic coupling and discussing its relation to molecular deformations.

2.2 Molecular Geometry and Modes of Motion

In this section, we discuss modes of a molecular motion taking a diatomic molecule, AB as an example.

Let us consider a molecule consisting of two atoms A and B. The masses are m_A and m_B, and the positions are denoted by (X_A, Y_A, Z_A) and (X_B, Y_B, Z_B), respectively.

According to classical mechanics, the kinetic energy T is given by, in the space-fixed Cartesian coordinate,

$$T = \frac{1}{2}m_A\left(\dot{X}_A^2 + \dot{Y}_A^2 + \dot{Z}_A^2\right) + \frac{1}{2}m_B\left(\dot{X}_B^2 + \dot{Y}_B^2 + \dot{Z}_B^2\right), \tag{2.1}$$

where the dots denote the derivative with respect to time t. Using the coordinates of the center of mass:

$$\begin{cases} X = \frac{m_A X_A + m_B X_B}{m_A + m_B} \\ Y = \frac{m_A Y_A + m_B Y_B}{m_A + m_B} \\ Z = \frac{m_A Z_A + m_B Z_B}{m_A + m_B} \end{cases}, \tag{2.2}$$

and the relative coordinates:

$$\begin{cases} x = X_A - X_B \\ y = Y_A - Y_B \\ z = Z_A - Z_B \end{cases}, \tag{2.3}$$

Equation (2.1) can be written as

$$T = \frac{1}{2}M\left(\dot{X}^2 + \dot{Y}^2 + \dot{Z}^2\right) + \frac{1}{2}m\left(\dot{x}^2 + \dot{y}^2 + \dot{z}^2\right), \tag{2.4}$$

where

$$M = m_A + m_B, \quad m = \frac{m_A m_B}{m_A + m_B}, \tag{2.5}$$

M is the total mass, and m is called a reduced mass. The first term of Eq. (2.4) describes the motion of the center of mass, and the second term the relative motion. Therefore, the motion of the center of mass is separated from the others.

Using the polar coordinate

$$\begin{aligned} x &= r \sin\theta \cos\phi \\ y &= r \sin\theta \sin\phi \ , \\ z &= r \cos\theta \end{aligned} \tag{2.6}$$

the second term of Eq. (2.4) is written as

$$T = \frac{1}{2}M\left(\dot{X}^2 + \dot{Y}^2 + \dot{Z}^2\right) + \frac{1}{2}m\dot{r}^2 + \frac{1}{2}m\left(r^2\dot{\theta}^2 + r^2\dot{\phi}^2 \sin^2\theta\right). \tag{2.7}$$

Using angular momentum

$$L = r^2\dot{\theta}, \tag{2.8}$$

the kinetic energy of a diatomic molecule is obtained as

$$T = \frac{1}{2}M\left(\dot{X}^2 + \dot{Y}^2 + \dot{Z}^2\right) + \frac{\mathbf{L}^2}{2mr^2} + \frac{1}{2}m\dot{r}^2. \tag{2.9}$$

As we will see later, the first term corresponds to a translation, the second term a rotation, and the third term a vibration, respectively. These are called modes of motion.

When we regard the molecule as a particle without the internal structure, neglecting (r, θ, ϕ), the kinetic energy can be written as

$$T = \frac{1}{2}M\left(\dot{X}^2 + \dot{Y}^2 + \dot{Z}^2\right) = \frac{1}{2M}\left(P_X^2 + P_Y^2 + P_Z^2\right) = \frac{\mathbf{P}^2}{2M}. \tag{2.10}$$

This is the kinetic energy of a single particle with a mass M which is the total sum of masses. When we consider the internal structure of the molecule, we can neglect the time dependence of the interatomic distance among the internal coordinate (r, θ, ϕ). Within this approximation, we obtain

$$T = \frac{\mathbf{P}^2}{2M} + \frac{\mathbf{L}^2}{2I}, \tag{2.11}$$

where $I := mr_0^2$ is called moment of inertia. The second term, Eq. (2.11), is the rotational energy of the rigid rotor. A moment of inertia in the rotational energy is like the mass in the translational energy as angular momentum in rotation is like momentum in translation:

$$\begin{array}{c} \mathbf{P} \leftrightarrow \mathbf{L} \\ M \leftrightarrow I \end{array} . \tag{2.12}$$

Equation (2.11) is the sum of the kinetic energies of the single particle and the rigid rotor.

Furthermore, if we take the derivative of r with respect to time, $\dot{r}$, into consideration, the kinetic energy can be written as

$$T = \frac{\mathbf{P}^2}{2M} + \frac{\mathbf{L}^2}{2I} + \frac{1}{2}m\dot{r}^2 = \frac{\mathbf{P}^2}{2M} + \frac{\mathbf{L}^2}{2I} + \frac{p_r^2}{2m}, \tag{2.13}$$

where $p_r = m\dot{r}$. The denominator in the second term of Eq. (2.9) is approximated by the constant $2I$. This approximation enables us to separate the rotation and the vibration of the kinetic energy which is described by the third term. Exactly speaking, I is not a constant, and the violation of the approximation can be observed if the molecule rotates, that is to say, the angular momentum is not equal to zero. The violation of this approximation is called effect of centrifugal deviation. However, this approximation is valid in most cases. Therefore, we can discuss the translation (X, Y, Z), rotation (θ, ϕ), and vibration r separately. Hereafter, we assume this approximation and concentrate ourselves on vibrations, or deformation of a molecule.

It is not easy to define a molecule. However, we can assume that a molecule has some stationary molecular structure: there exists a certain interatomic distance $r = r_0$ such that

$$\left(\frac{\partial E(r)}{\partial r}\right)_{r_0} = 0 \quad \text{and} \quad \left(\frac{\partial^2 E(r)}{\partial r^2}\right)_{r_0} > 0, \tag{2.14}$$

where $E(r)$ is the potential energy. This is because we can expect that interactions among electrons and nuclei could yield such an attracting potential. The Taylor expansion of $E(r)$ around $r = r_0$ is expressed by, using Eq. (2.14),

$$\begin{aligned} E(r) &= E_0 + \left(\frac{\partial E(r)}{\partial r}\right)_{r_0} (r - r_0) + \frac{1}{2!}\left(\frac{\partial^2 E(r)}{\partial r^2}\right)_{r_0} (r - r_0)^2 + \cdots \\ &= E_0 + \frac{1}{2}k(r - r_0)^2 + \cdots , \end{aligned} \tag{2.15}$$

where $E_0 = E(r_0)$ and

$$k = \left(\frac{\partial^2 E(r)}{\partial r^2}\right)_{r_0} \tag{2.16}$$

is called a force constant.

Accordingly, we can obtain the following approximate molecular Hamiltonian:

$$\hat{H} = \frac{\hat{\boldsymbol{P}}^2}{2M} + \frac{\hat{\boldsymbol{L}}^2}{2I} + \frac{p_r^2}{2m} + E_0 + \frac{1}{2}k(r - r_0)^2 = E_0 + \hat{H}_{\mathrm{t}} + \hat{H}_{\mathrm{r}} + \hat{H}_{\mathrm{v}}, \tag{2.17}$$

where

$$\hat{H}_{\mathrm{t}} = \frac{\hat{\boldsymbol{P}}^2}{2M} \tag{2.18}$$

is the Hamiltonian of the translational modes,

$$\hat{H}_{\mathrm{r}} = \frac{\hat{\boldsymbol{L}}^2}{2I} \tag{2.19}$$

is the Hamiltonian of the rotational modes, and

$$\hat{H}_{\mathrm{v}} = \frac{p_r^2}{2m} + \frac{1}{2}k(r - r_0)^2 \tag{2.20}$$

is the Hamiltonian of the vibrational mode.

We separate the translation and rotation modes from the molecular motion hereafter since internal molecular motions or deformation can be regarded as a vibration as shown in Fig. 2.1. In the case of a diatomic molecule AB, the intramolecular structure is specified by the interatomic distance r. The vibrational Hamiltonian after the separation is given by

$$\hat{H}_{\mathrm{v}} = \frac{1}{2m}\hat{P}_r^2 + \frac{1}{2}k(r - r_0)^2 = \frac{1}{2m}\hat{P}_q^2 + \frac{1}{2}kq^2, \tag{2.21}$$

where q is the displacement or deformation

$$q = r - r_0. \tag{2.22}$$

Note that a molecular deformation is described by vibrational coordinates.

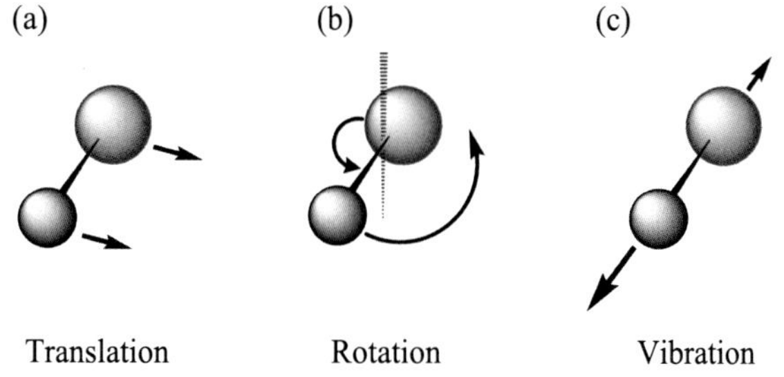

Fig. 2.1 Freedom of a diatomic molecule, translation, rotation, and vibration

2.3 Coordinates of Non-linear A_3

In this section, we discuss the internal coordinates of triatomic molecules, A_3. The molecular structure can be linear or non-linear. When it comes to a molecular structure, we implicitly consider the molecule as a rigid body. It is important to distinguish a linear or non-linear molecular structure since the degree of freedom is different between them. The rotation of a linear molecule is described by (θ, ϕ) as the same as a diatomic molecule, and that of a non-linear molecule is, for example, by Euler angle (α, β, γ). Therefore, the rotational degree of freedom is two and three for a linear and non-linear molecules, respectively. Since the total degree of freedom is $3 \times 3 = 9$ and the translational degree of freedom is three, the remaining vibrational degree of freedom is four and three, respectively. A molecular deformation is described by four or three vibrational coordinates depending on its molecular structure as a rigid body.

In this section, we concentrate ourselves on a non-linear A_3. The largest point group of non-linear A_3 is D_{3h}. The Cartesian coordinates are denoted by

$$\begin{array}{l} A_1(X_1, Y_1, 0) \\ A_2(X_2, Y_2, 0), \\ A_3(X_3, Y_3, 0) \end{array} \tag{2.23}$$

where the molecule is placed on the xy-plane. We will derive symmetry-adapted vibrational coordinates of the D_{3h}-A_3 (Fig. 2.2).

Since the shape of the molecules is a triangle, we can adopt the three bond lengths (see Fig. 2.3),

$$l_i = \sqrt{(X_{i+1} - X_i)^2 + (Y_{i+1} - Y_i)^2}, \quad (i = 1, 2, 3) \tag{2.24}$$

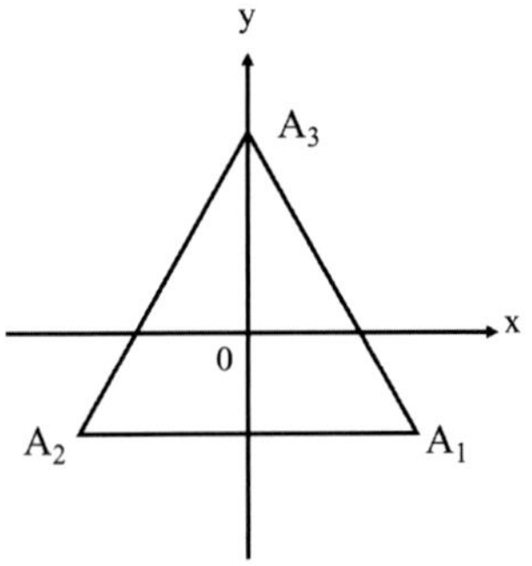

Fig. 2.2 Equilateral triangle (C_{3v})

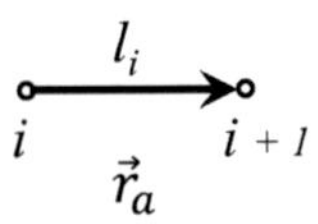

Fig. 2.3 Bond length as an internal coordinate

as internal (vibrational) coordinates. The Cartesian displacements from the equilibrium position $(X_i^0, Y_i^0, 0)$ are written as

$$x_i = X_i - X_i^0, \quad y_i = Y_i - Y_i^0. \tag{2.25}$$

A change in bond length l_i is given by

$$\Delta l_i = l_i - l_i^0. \tag{2.26}$$

The bond length is, by expanding for $x_i, x_{i+1}, y_i, y_{i+1}$ around the equilibrium position,

$$l_i = \sqrt{(X_{i+1}^0 + x_{i+1} - X_i^0 - x_i)^2 + (Y_{i+1}^0 + y_{i+1} - Y_i^0 - y_i)^2}, \tag{2.27}$$

$$\approx l_i^0 + \frac{1}{l_i^0}\left[(X_{i+1}^0 - X_i^0)(x_{i+1} - x_i) + (Y_{i+1}^0 - Y_i^0)(y_{i+1} - y_i)\right], \tag{2.28}$$

where l_i^0 is the equilibrium bond length. Therefore, the change of the bond length is expressed by

$$\Delta l_i = l_i - l_i^0 = \frac{1}{l_i^0}\left[(X_{i+1}^0 - X_i^0)(x_{i+1} - x_i) + (Y_{i+1}^0 - Y_i^0)(y_{i+1} - y_i)\right]. \tag{2.29}$$

For D_{3h}-A3 (see Fig. 2.2), all the bond lengths are equal: $l_1^0 = l_2^0 = l_3^0 = l_0$, and the Cartesian coordinates can be taken as

$$\mathrm{A}_1\left(\frac{l^0}{2}, -\frac{l^0}{2\sqrt{3}}\right), \quad \mathrm{A}_2\left(-\frac{l^0}{2}, -\frac{l^0}{2\sqrt{3}}\right), \quad \mathrm{A}_3\left(0, \frac{l^0}{\sqrt{3}}\right). \tag{2.30}$$

Accordingly, the changes of the bond lengths are obtained using (2.29) as

$$\Delta l_1 = -(x_2 - x_1) \tag{2.31}$$

$$\Delta l_2 = \frac{1}{2}(x_3 - x_2) + \frac{\sqrt{3}}{2}(y_3 - y_2) \tag{2.32}$$

$$\Delta l_3 = \frac{1}{2}(x_1 - x_3) - \frac{\sqrt{3}}{2}(y_1 - y_3). \tag{2.33}$$

These bond length changes describe bond stretching motions.

We consider the present problem under C_{3v} which is a subgroup of D_{3h}. The rotation about the threefold axis C_3 shown in Fig. 2.4 is written as

$$\boldsymbol{x}' = \hat{C}_3\boldsymbol{x} = \begin{pmatrix} x' \\ y' \end{pmatrix} = \begin{pmatrix} \cos\frac{2\pi}{3} & -\sin\frac{2\pi}{3} \\ \sin\frac{2\pi}{3} & \cos\frac{2\pi}{3} \end{pmatrix}\begin{pmatrix} x \\ y \end{pmatrix} = \begin{pmatrix} -\frac{1}{2} & -\frac{\sqrt{3}}{2} \\ \frac{\sqrt{3}}{2} & -\frac{1}{2} \end{pmatrix}\begin{pmatrix} x \\ y \end{pmatrix}, \tag{2.34}$$

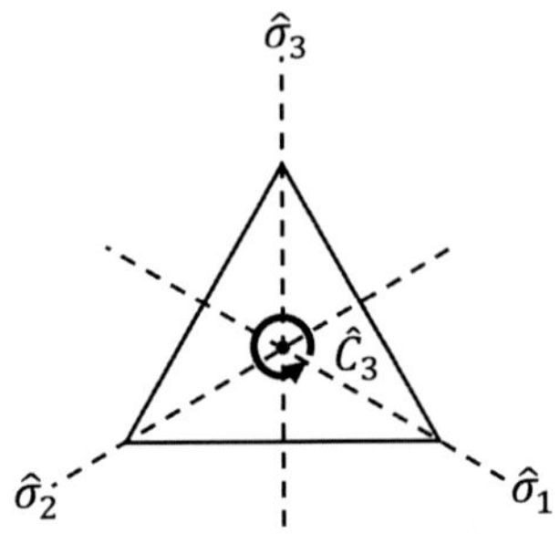

Fig. 2.4 Symmetry operations of C_{3v}

where

$$\boldsymbol{x} = (x, y)^{\mathrm{T}} = \begin{pmatrix} x \\ y \end{pmatrix} \tag{2.35}$$

is the Cartesian coordinate of an arbitrary point on the 2D-plane. The reflection against σ_3 is written as

$$\boldsymbol{x}' = \hat{\sigma}_3 \boldsymbol{x} = \begin{pmatrix} x' \\ y' \end{pmatrix} = \begin{pmatrix} -1 & 0 \\ 0 & 1 \end{pmatrix} \begin{pmatrix} x \\ y \end{pmatrix}. \tag{2.36}$$

The atomic position vector $\boldsymbol{R}_i = (X_i, Y_i)^{\mathrm{T}}$ of a deformed molecule with a deformation vector $\boldsymbol{x}_i = (x_i, y_i)^{\mathrm{T}}$ is defined by

$$\boldsymbol{R}_i = \boldsymbol{R}_i^0 + \boldsymbol{x}_i, \tag{2.37}$$

where $\boldsymbol{R}_i^0 = (X_i^0, Y_i^0)^{\mathrm{T}}$ is the atomic position vector of a molecule at the reference geometry $\boldsymbol{R}^0$. The atomic positions at the reference geometry are transformed as

$$\hat{C}_3 : A_1 \mapsto A_2 \mapsto A_3, \tag{2.38}$$

and

$$\hat{\sigma}_3 : A_1 \leftrightarrow A_2, \quad A_3 \mapsto A_3. \tag{2.39}$$

We will employ group theory hereafter. See Ref. [1] for the definitions and theorems. We can take $\hat{\sigma}_3$ and $\hat{C}_3$ as generators[1] of the point group C_{3v}. We can find by inspection that the actions of these generators on Δl_i are as follows:

$$\begin{aligned} \hat{\sigma}_3 \Delta l_1 &= \Delta l_1 \\ \hat{\sigma}_3 \Delta l_2 &= \Delta l_3 \,, \\ \hat{\sigma}_3 \Delta l_3 &= \Delta l_2 \end{aligned} \tag{2.40}$$

[1] The generators generate group by multiplying themselves and their products.

and

$$\begin{aligned}\hat{C}_3\Delta l_1 &= \Delta l_3\\ \hat{C}_3\Delta l_2 &= \Delta l_1\,.\\ \hat{C}_3\Delta l_3 &= \Delta l_2\end{aligned} \tag{2.41}$$

Therefore, the representations[2] of these generators in $(\Delta l_1, \Delta l_2, \Delta l_3)$ are obtained as

$$D_\Gamma(\hat{\sigma}_3) = \begin{pmatrix} 1 & 0 & 0\\ 0 & 0 & 1\\ 0 & 1 & 0\end{pmatrix}, \quad D_\Gamma(\hat{C}_3) = \begin{pmatrix} 0 & 0 & 1\\ 1 & 0 & 0\\ 0 & 1 & 0\end{pmatrix}. \tag{2.42}$$

Using (2.42), we can derive the other representation matrices

$$D_\Gamma(\hat{C}_3^2) = D_\Gamma(\hat{C}_3)D_\Gamma(\hat{C}_3) = \begin{pmatrix} 0 & 1 & 0\\ 0 & 0 & 1\\ 1 & 0 & 0\end{pmatrix}, \tag{2.43}$$

$$D_\Gamma(\hat{\sigma}_1) = D_\Gamma(\hat{C}_3^2\sigma_3) = D_\Gamma(\hat{C}_3^2)D_\Gamma(\sigma_3) = \begin{pmatrix} 0 & 0 & 1\\ 0 & 1 & 0\\ 1 & 0 & 0\end{pmatrix}, \tag{2.44}$$

and

$$D_\Gamma(\hat{\sigma}_2) = D_\Gamma(\hat{C}_3\sigma_3) = D_\Gamma(\hat{C}_3)D_\Gamma(\sigma_3) = \begin{pmatrix} 0 & 1 & 0\\ 1 & 0 & 0\\ 0 & 0 & 1\end{pmatrix}. \tag{2.45}$$

Using these matrices, we obtain the characters $\chi_\Gamma(R)$[3] of the representation matrices as tabulated in Table 2.1 as well as the irreps[4] of C_{3v}.

By inspection from Table 2.1, we obtain the irreducible decomposition of Γ:

$$\Gamma = A_1 + E. \tag{2.46}$$

Using Δl_i, we can describe the totally-symmetric A_1 deformation which is invariant under the symmetry operations in D_{3h} as follows:

$$q_{A_1} = \frac{1}{3}(\Delta l_1 + \Delta l_2 + \Delta l_3) = \frac{1}{2}x_1 - \frac{1}{2\sqrt{3}}y_1 - \frac{1}{2}x_2 - \frac{1}{2\sqrt{3}}y_2 + 0x_3 + \frac{1}{\sqrt{3}}y_3. \tag{2.47}$$

---

[2]They represent the symmetry operation as the matrices on the basis of a certain basis, Δl_i in the present case.

[3]Character is the trace of a representation matrix, $\chi_\Gamma(R) = \mathrm{tr}D_\Gamma(\mathrm{R})$.

[4]irrep is an acronym of irreducible representation. See Ref. [1].

Table 2.1 Characters of Γ and the irreducible representations of C_{3v}

	$\hat{E}$	$2\hat{C}_3$	$3\hat{\sigma}_3$
Γ	3	0	1
A_1	1	1	1
E	2	-1	0

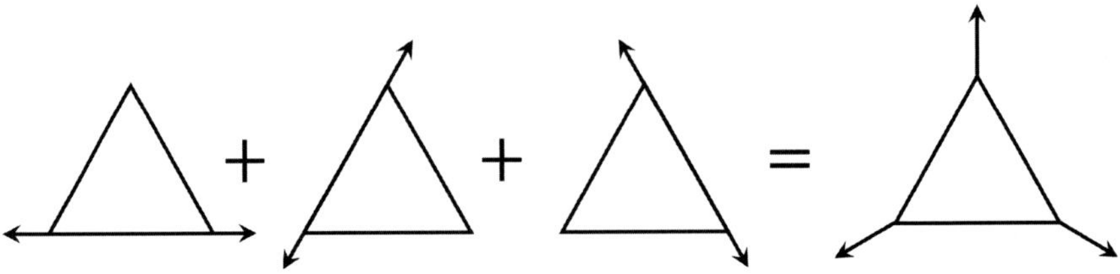

Fig. 2.5 Totally-symmetric deformation of the equilateral triangle described by q_{A_1}

Table 2.2 Irreducible representation matrices of C_{3v}

C_{3v}	$\hat{E}$	$\hat{C}_3$	$\hat{C}_3^2$	$\hat{\sigma}_1$	$\hat{\sigma}_2$	$\hat{\sigma}_3$
A_1	1	1	1	1	1	1
A_2	1	1	1	-1	-1	-1
E	$\begin{pmatrix} 1 & 0 \\ 0 & 1 \end{pmatrix}$	$\begin{pmatrix} -\frac{1}{2} & -\frac{\sqrt{3}}{2} \\ \frac{\sqrt{3}}{2} & -\frac{1}{2} \end{pmatrix}$	$\begin{pmatrix} -\frac{1}{2} & \frac{\sqrt{3}}{2} \\ -\frac{\sqrt{3}}{2} & -\frac{1}{2} \end{pmatrix}$	$\begin{pmatrix} 1 & 0 \\ 0 & -1 \end{pmatrix}$	$\begin{pmatrix} -\frac{1}{2} & -\frac{\sqrt{3}}{2} \\ -\frac{\sqrt{3}}{2} & \frac{1}{2} \end{pmatrix}$	$\begin{pmatrix} -\frac{1}{2} & \frac{\sqrt{3}}{2} \\ \frac{\sqrt{3}}{2} & \frac{1}{2} \end{pmatrix}$

Under any deformation along q_{A_1}, the molecule keeps its shape a equilateral triangle as shown in Fig. 2.5. The point group of the deformed molecules is D_{3h}. A symmetry-adapted coordinate such as q_{A_1} can be obtained using projection operators [1]:

$$\hat{P}_{kl}^{\Omega} = \frac{\dim(\Omega)}{|G|} \sum_{R \in G} \bar{D}_{kl}^{\Omega}(R) \hat{R}, \tag{2.48}$$

where $|G|$ denotes the number of elements (order) in group G, $\dim(\Omega)$ the dimension of irrep Ω, and $\bar{D}_{kl}^{\Omega}(R)$ the complex conjugate of (k, l) element of the irreducible representation matrix of symmetry operation R in irrep Ω (Table 2.2).

For A_1 representation, we obtain

$$\hat{P}^{A_1} = \frac{1}{6}\left(1 \cdot \hat{E} + 1 \cdot \hat{C}_3 + 1 \cdot \hat{C}_3^2 + 1 \cdot \hat{\sigma}_1 + 1 \cdot \hat{\sigma}_2 + 1 \cdot \hat{\sigma}_3\right). \tag{2.49}$$

When we take Δl_1 and apply $\hat{P}^{A_1}$, we can derive the expression of q_{A_1}. There exist doubly degenerate E coordinates, $q_{E\theta}$ and $q_{E\epsilon}$, in D_{3h}-A_3. The projection operators for $q_{E\theta}$ and $q_{E\epsilon}$ are

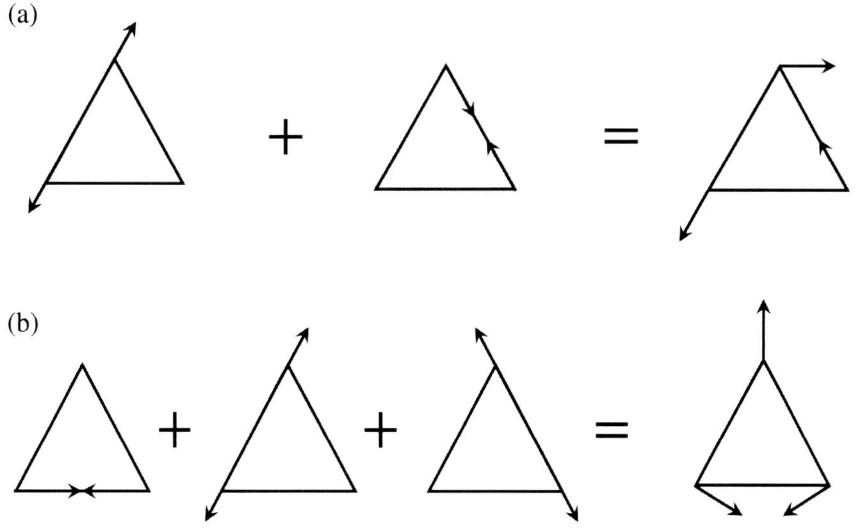

Fig. 2.6 Degenerate deformations of the equilateral triangle described by **a** $q_{E\epsilon}$ and **b** $q_{E\theta}$

$$\hat{P}_{11}^{E\theta} = \frac{2}{6}\left(1 \cdot \hat{E} + (-\frac{1}{2}) \cdot \hat{C}_3 + (-\frac{1}{2}) \cdot \hat{C}_3^2 + (\frac{1}{2}) \cdot \hat{\sigma}_1 + (\frac{1}{2}) \cdot \hat{\sigma}_2 + (-1) \cdot \hat{\sigma}_3\right) \tag{2.50}$$

and

$$\hat{P}_{22}^{E} = \frac{2}{6}\left(1 \cdot \hat{E} + (-\frac{1}{2}) \cdot \hat{C}_3 + (-\frac{1}{2}) \cdot \hat{C}_3^2 + (-\frac{1}{2}) \cdot \hat{\sigma}_1 + (-\frac{1}{2}) \cdot \hat{\sigma}_2 + 1 \cdot \hat{\sigma}_3\right), \tag{2.51}$$

respectively. We can obtain $q_{E\theta}$ and $q_{E\epsilon}$ by applying $\hat{P}_{11}^{E\theta}$ and $\hat{P}_{22}^{E}$ for Δl_2 (note that $q_{E\theta}$ does not contain Δl_1):

$$\hat{P}_{11}^{E} \Delta l_2 = \frac{1}{2}(\Delta l_2 - \Delta l_3) =: q_{E\epsilon}, \tag{2.52}$$

the three bond lengths become different. The point group of the deformed molecules is C_1 as shown in Fig. 2.6a. On the other hand, along

$$\hat{P}_{22}^{E} \Delta l_2 = \frac{1}{6}(-2\Delta l_1 + \Delta l_2 + \Delta l_3) =: q_{E\theta}. \tag{2.53}$$

As shown in Fig. 2.6b, the molecule deforms into an isosceles triangle along $q_{E\theta}$. The point group of the deformed molecules is C_{2v}.

In summary, the symmetry-adapted vibrational coordinates are expressed by internal coordinates, Δl_1, Δl_2, Δl_3 as

$$
\begin{aligned}
q_{A_1} &= \frac{1}{3}\left(\Delta l_1 + \Delta l_2 + \Delta l_3\right), \\
q_{E\epsilon} &= \frac{1}{2}(\Delta l_2 - \Delta l_3), \\
q_{E\theta} &= \frac{1}{6}(-2\Delta l_1 + \Delta l_2 + \Delta l_3).
\end{aligned} \tag{2.54}
$$

A molecular deformation of D_{3h}-A_3 is described by these coordinates.

2.4 Coordinates of Non-linear A_2B

In the case of A_3, the vibrational coordinates can be obtained by their symmetry because the representations in terms of $(\Delta l_1, \Delta l_2, \Delta l_3)$ do not contain any equivalent irreducible representation. The largest point group of non-linear A_2B is C_{2v} (Fig. 2.7). The symmetry operations of C_{2v} are depicted in Fig. 2.8. The characters of the irreps are tabulated in Table 5.5.

As discussed in Sect. 2.3, we obtain the representation matrices as

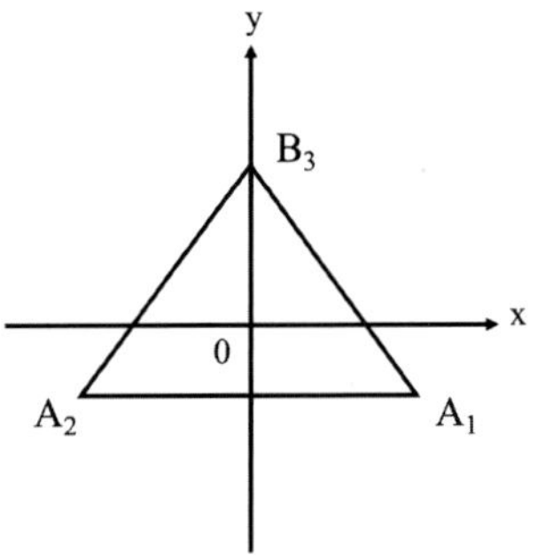

Fig. 2.7 Isosceles triangle (C_{2v})

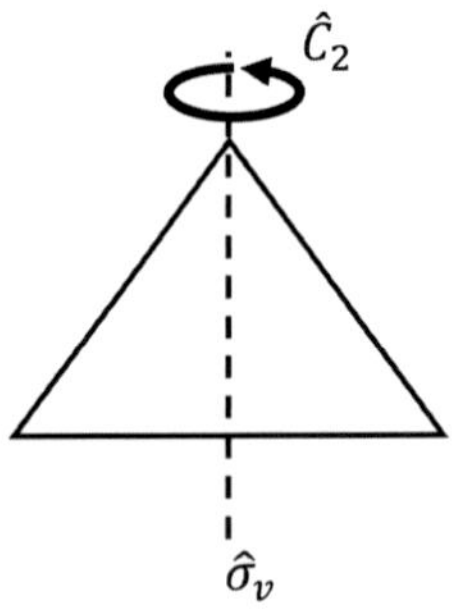

Fig. 2.8 Symmetry operations of C_{2v}

Table 2.3 Characters of Γ_{A_2B} and the irreducible representations of C_{2v}

	$\hat{E}$	$\hat{C}_2$	$\hat{\sigma}(xz)$	$\hat{\sigma}(yz)$
Γ_{A_2B}	3	1	3	1
A_1	1	1	1	1
A_2	1	1	−1	−1
B_1	1	−1	1	−1
B_2	1	−1	−1	1

$$D(\hat{C}_2) = \begin{pmatrix} 1 & 0 & 0 \\ 0 & 0 & 1 \\ 0 & 1 & 0 \end{pmatrix}, \quad D(\hat{\sigma}(yz)) = \begin{pmatrix} 1 & 0 & 0 \\ 0 & 0 & 1 \\ 0 & 1 & 0 \end{pmatrix}, \quad D(\hat{\sigma}(xz)) = \begin{pmatrix} 1 & 0 & 0 \\ 0 & 1 & 0 \\ 0 & 0 & 1 \end{pmatrix}. \tag{2.55}$$

The characters in this representation, Γ_{A_2B}, are tabulated in Table 2.3.

By inspection, the representation can be decomposed into

$$\Gamma_{A_2B} = 2A_1 + B_1. \tag{2.56}$$

Therefore, the symmetry-adapted coordinate for B_1 can be obtained only from symmetry consideration. The projection operators are derived as

$$\begin{aligned} \hat{P}^{A_1} &= \frac{1}{4}(\hat{E} + \hat{C}_2 + \hat{\sigma}(xz) + \hat{\sigma}(yz)), \\ \hat{P}^{B_1} &= \frac{1}{4}(\hat{E} - \hat{C}_2 + \hat{\sigma}(xz) - \hat{\sigma}(yz)). \end{aligned} \tag{2.57}$$

Applying $\hat{P}^{B_1}$ for $\Delta l_2 - \Delta l_3$, we can obtain

$$q_{B_1} := \hat{P}^{B_1}(\Delta l_2 - \Delta l_3) = \Delta l_2 - \Delta l_3. \tag{2.58}$$

The symmetry-adapted coordinate q_{B_1} is one of the normal coordinates of A_2B, but the normal coordinates for representation A_1 cannot be determined only from symmetry consideration. Applying $\hat{P}^{A_1}$ for $\Delta l_1 + \Delta l_2 + \Delta l_3$ and $-2\Delta l_1 + \Delta l_2 + \Delta l_3$, we can obtain the following two independent symmetry-adapted coordinates:

$$\begin{aligned} q_{A_1 a} &:= \hat{P}^{A_1}(\Delta l_2 + \Delta l_2 + \Delta l_3) = \Delta l_2 + \Delta l_2 + \Delta l_3, \\ q_{A_1 b} &:= \hat{P}^{A_1}(-2\Delta l_1 + \Delta l_2 + \Delta l_3) = -2\Delta l_1 + \Delta l_2 + \Delta l_3. \end{aligned} \tag{2.59}$$

The normal coordinates for representation A_1 can be obtained as a linear combination of $q_{A_1 a}$ and $q_{A_1 b}$ from the vibrational analysis.

2.5 Approximate Molecular Hamiltonian

For a polyatomic molecule, one way to specify the molecular structure is to use bond angles and dihedral angles as well as bond lengths as internal coordinates. A bond angle φ_i (see Fig. 2.9) is defined by

$$\varphi_i = \arccos\left(\frac{\vec{r}_\mathrm{a} \cdot \vec{r}_\mathrm{b}}{r_\mathrm{a} r_\mathrm{b}}\right), \tag{2.60}$$

and a dihedral angle χ_i is defined by the angle between the two planes which are spanned by r_a and r_b, and r_b and r_c (see Fig. 2.10). The dihedral angle between two planes is given by the angle between the vectors normal to these planes:

$$\chi_i = \arccos\left(\frac{\vec{r}_\mathrm{a} \cdot (\vec{r}_\mathrm{b} \times \vec{r}_\mathrm{c})}{r_a r_b r_c}\right). \tag{2.61}$$

A set of bond lengths, bond angles, and dihedral angles to specify a molecular structure is called Z-matrix. To derive a quantum-mechanical Hamiltonian using such internal coordinates is difficult since the canonical quantization condition is only valid for Cartesian coordinates. As the same in a bond length change Δl_i, the changes of a bond angle $\Delta\varphi_i$ and dihedral angle $\Delta\chi_i$ can be expressed as a linear combination of Cartesian coordinates by expanding them around the equilibrium position. We can obtain an approximate molecular Hamiltonian by expanding potential energy in terms of Cartesian displacement coordinates up to the second order. The linear terms are vanishing for an equilibrium structure, and the quadratic terms can be expressed as a standard form via vibration analysis. In the vibrational analysis using Cartesian displacement coordinates, normal modes with finite eigenvalues correspond to vibrational modes. Finally, collecting these normal modes, the vibrational Hamiltonian is expressed by, within the harmonic approximation,

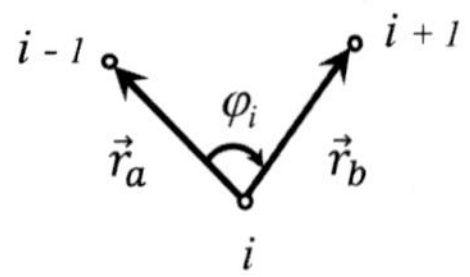

Fig. 2.9 Bond angle as an internal coordinate

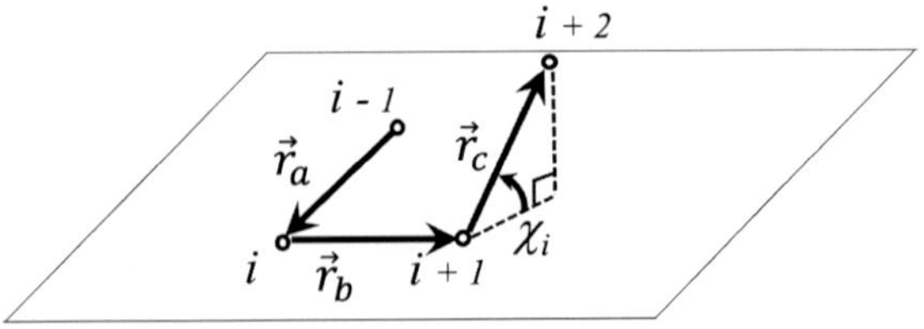

Fig. 2.10 Dihedral angle as an internal coordinate

$$\hat{H}_{\mathrm{v}} = \sum_{\alpha}\left[-\frac{\hbar^2}{2\mu_\alpha}\frac{\partial^2}{\partial q_\alpha^2} + \frac{1}{2}K_\alpha q_\alpha^2\right], \tag{2.62}$$

where q_α is called normal coordinate, and μ_α is the reduced mass of mode α. As discussed in Sect. 5.3, the Hamiltonian can be simplified by introducing mass-weighted coordinate

$$Q_\alpha = \sqrt{\mu_\alpha} q_\alpha, \tag{2.63}$$

as

$$\hat{H}_{\mathrm{v}} = \sum_{\alpha}\left[-\frac{\hbar^2}{2}\frac{\partial^2}{\partial Q_\alpha^2} + \frac{1}{2}\omega_\alpha^2 Q_\alpha^2\right], \tag{2.64}$$

where

$$K_\alpha = \mu_\alpha \omega_\alpha^2. \tag{2.65}$$

We will denote a molecular deformation by a mass-weighted coordinate hereafter.

2.6 Energy as a Function of Molecular Deformation

We have defined in Sect. 2.1 the reference geometry as a stable structure of a certain electronic state $|\Psi_0\rangle$ or an assumed geometry such as planar NH_3 and linear H_2O discussed in Chap. 1. We call $|\Psi_0\rangle$ a reference state. A molecular deformation from a reference structure $\boldsymbol{R}^0 = (X_i^0, Y_i^0, Z_i^0)$ can be expressed by a set of internal coordinate, Δl_i, $\Delta\varphi_i$, and $\Delta\chi_i$. This set transforms into the set of the normal coordinates, Q_α. We describe a molecular deformation in terms of Q_α hereafter. Accordingly, any deformed structure $\boldsymbol{R} = (X_i, Y_i, Z_i)$ is written as

$$\boldsymbol{R} = \boldsymbol{R}^0 + \Delta\boldsymbol{R} = \boldsymbol{R}^0 + \Delta\boldsymbol{R}(\{\Delta l_i, \Delta\varphi_i, \Delta\chi_i\}) = \boldsymbol{R}^0 + \Delta\boldsymbol{R}(\{Q_\alpha\}) = \boldsymbol{R}^0 + \Delta\boldsymbol{R}(\boldsymbol{Q}), \tag{2.66}$$

where the normal coordinate is defined for the reference state, $|\Psi_0\rangle$, at the reference geometry, $\boldsymbol{R}^0$. Therefore, the Hessian

$$\left(\frac{\partial^2 E_0}{\partial Q_\alpha \partial Q_\beta}\right)_{\boldsymbol{R}^0} \tag{2.67}$$

is diagonal. Note that the diagonal elements can be positive and negative. As we have discussed a molecular deformation, taking A_3 and A_2B as examples in the previous sections, a molecular deformation $\Delta\boldsymbol{R}$ can be expressed in terms of mass-weighted normal coordinates as follows (see Sect. 5.3):

$$\Delta\boldsymbol{R}(\boldsymbol{Q}) = \sum_{\alpha} Q_\alpha \boldsymbol{u}_\alpha. \tag{2.68}$$

The energy $E(\boldsymbol{R})$ can be expanded around the reference geometry $\boldsymbol{R}^0$,

$$E(\boldsymbol{R}) = E(\boldsymbol{R}^0) + \left(\frac{\partial E}{\partial \Delta \boldsymbol{R}}\right)_{\boldsymbol{R}^0} \cdot \Delta \boldsymbol{R} + \frac{1}{2}\Delta \boldsymbol{R}^{\mathrm{T}} \left(\frac{\partial^2 E}{\partial \Delta \boldsymbol{R} \partial \Delta \boldsymbol{R}}\right)_{\boldsymbol{R}^0} \Delta \boldsymbol{R} + \cdots \tag{2.69}$$

$$= E(\boldsymbol{R}^0) + \sum_{\alpha} \left(\frac{\partial E}{\partial Q_\alpha}\right)_{\boldsymbol{R}^0} Q_\alpha + \frac{1}{2}\sum_{\alpha,\beta} \left(\frac{\partial^2 E}{\partial Q_\alpha \partial Q_\beta}\right)_{\boldsymbol{R}^0} Q_\alpha Q_\beta + \cdots \tag{2.70}$$

The equilibrium point $\boldsymbol{R}^0$ is defined by the following equation:

$$\left(\frac{\partial E_{\mathrm{ref}}(\boldsymbol{R})}{\partial Q_\alpha}\right)_{\boldsymbol{R}^0} = 0, \tag{2.71}$$

where E_{ref} is the electronic energy for the reference electronic state $|\Psi_{\mathrm{ref}}\rangle$ defined in the next section.

2.7 Molecular Deformation Due to Vibronic Couplings

We start from the Herzberg–Teller expansion (see Sect. 5.4.1):

$$\hat{H}(\boldsymbol{R}) = \hat{T}_{\mathrm{n}}(\boldsymbol{Q}) + \hat{H}_{\mathrm{e}}(\boldsymbol{R}^0) + \sum_{\alpha} \left(\frac{\partial \hat{H}}{\partial Q_\alpha}\right)_{\mathbf{R}^0} Q_\alpha + \frac{1}{2}\sum_{\alpha,\beta} \left(\frac{\partial^2 \hat{H}}{\partial Q_\alpha \partial Q_\beta}\right)_{\mathbf{R}^0} Q_\alpha Q_\beta + \cdots . \tag{2.72}$$

The second term is called linear vibronic coupling, and the third term quadratic vibronic coupling.

For vibronic wave function $\Phi(\boldsymbol{r}, \boldsymbol{R})$, we employ the crude adiabatic representation:

$$\Phi(\boldsymbol{r}, \boldsymbol{R}) = \sum_{n} \Psi_n(\boldsymbol{r}, \boldsymbol{R}^0)\chi_{ni}(\boldsymbol{R}), \tag{2.73}$$

where $\boldsymbol{r}$ is a set of N-electron coordinates, $\Psi_n(\boldsymbol{r}, \boldsymbol{R}^0)$ is an electronic state at a reference geometry $\boldsymbol{R}^0$, and $\chi_{ni}(\boldsymbol{R})$ denotes nuclear wave function with vibrational quantum number i for electronic state Ψ_n. The electronic state satisfies

$$\hat{H}_{\mathrm{e}}(\boldsymbol{R}^0)\Psi_n(\boldsymbol{r}, \boldsymbol{R}^0) = E_n(\boldsymbol{R}^0)\Psi_n(\boldsymbol{r}, \boldsymbol{R}^0). \tag{2.74}$$

Using the basis set $\{\Psi_n(\boldsymbol{r}, \boldsymbol{R}^0)\}$, Hamiltonian (2.72) is represented as

$$(\hat{\mathbf{H}})_{mn} = \langle \Psi_m(\boldsymbol{r}, \boldsymbol{R}^0)|\hat{H}(\boldsymbol{R})|\Psi_n(\boldsymbol{r}, \boldsymbol{R}^0)\rangle \tag{2.75}$$

$$= \hat{T}_{\mathrm{n}}(\mathbf{Q})\delta_{mn} + E_n(\boldsymbol{R}^0)\delta_{mn} + \sum_{\alpha} V_{mn,\alpha} Q_\alpha + \frac{1}{2}\sum_{\alpha,\beta} W_{mn,\alpha\beta} Q_\alpha Q_\beta + \cdots , \tag{2.76}$$

where

$$V_{mn,\alpha} = \langle \Psi_m(\boldsymbol{r}, \boldsymbol{R}^0)| \left(\frac{\partial \hat{H}}{\partial Q_\alpha} \right)_{\mathbf{R}^0} |\Psi_n(\boldsymbol{r}, \boldsymbol{R}^0)\rangle \tag{2.77}$$

and

$$W_{mn,\alpha\beta} = \langle \Psi_m(\boldsymbol{r}, \boldsymbol{R}^0)| \left(\frac{\partial^2 \hat{H}}{\partial Q_\alpha \partial Q_\beta} \right)_{\mathbf{R}^0} |\Psi_n(\boldsymbol{r}, \boldsymbol{R}^0)\rangle. \tag{2.78}$$

Hamiltonian matrix (2.76) is called vibronic Hamiltonian. $V_{mn,\alpha}$ is called a linear vibronic coupling constant, which describes the electronic part of a linear vibronic coupling. $W_{mn,\alpha\beta}$ is called a quadratic vibronic coupling constant which describes the electronic part of a quadratic vibronic coupling.

The first term of Hamiltonian matrix (2.76) is nuclear kinetic energy matrix, and the remaining terms describe potential energy. When we neglect the nuclear kinetic energy, we can obtain a static picture of a molecule. A representation that is diagonal in nuclear kinetic energy is called a diabatic representation, and a representation that is diagonal in potential energy is called an adiabatic representation. In this respect, the vibronic Hamiltonian $\hat{\mathbf{H}}$ is a diabatic representation.

2.7.1 *Force Acting on Molecule: Linear Vibronic Couplings*

In this subsection, we assume electronic state n is not degenerate. When we neglect the off-diagonal elements of $V_{mn,\alpha}$ and $W_{mn,\alpha\beta}$ $(m \neq n)$, we obtain

$$(\hat{\mathbf{H}})_{nn} = \hat{T}_{\mathrm{n}}(\mathbf{Q}) + E_n(\boldsymbol{R}^0) + \sum_\alpha V_{nn,\alpha} Q_\alpha + \frac{1}{2} \sum_{\alpha,\beta} W_{nn,\alpha\beta} Q_\alpha Q_\beta + \cdots . \tag{2.79}$$

This approximation is called crude adiabatic approximation. If the normal modes for Ψ_n are not so different from those for Ψ_0, the cross terms in the third term of Eq. (2.79) can be neglected:

$$\begin{aligned}(\hat{\mathbf{H}})_{nn} &= \hat{T}_{\mathrm{n}}(\mathbf{Q}) + E_n(\boldsymbol{R}^0) + \sum_\alpha V_{nn,\alpha} Q_\alpha + \frac{1}{2} \sum_\alpha W_{nn,\alpha\alpha} Q_\alpha^2 + \cdots , \\ &= E_n(\boldsymbol{R}^0) + \sum_\alpha \left[-\frac{\hbar^2}{2} \frac{\partial^2}{\partial Q_\alpha} + V_{nn,\alpha} Q_\alpha + \frac{1}{2} W_{nn,\alpha\alpha} Q_\alpha^2 + \right] \cdots . \end{aligned} \tag{2.80}$$

The second term is the sum of the Hamiltonian for a displaced oscillator with frequency ω:

$$-\frac{\hbar^2}{2} \frac{\partial^2}{\partial Q^2} - |V| Q + \frac{1}{2} \omega^2 Q^2 = -\frac{\hbar^2}{2} \frac{\partial^2}{\partial Q^2} + \frac{1}{2} \omega^2 \left(Q - \frac{|V|}{\omega^2} \right)^2 - \frac{V^2}{2\omega^2}. \tag{2.81}$$

The appearance of the linear term in Q, $-|V|Q$, gives rise to the displacement of the minimum position from the origin by $Q = |V|/\omega^2$, and the minimum energy is lowered by $\Delta E = V^2/2\omega^2$. Here, since we can choose the direction of a vibrational vector, we define the direction for the system to be stabilized for a positive Q hereafter. The choice of the vibrational vectors results in negative VCCs, and thus we can write $V_\alpha = -|V_\alpha|$. Equation (2.80) is rewritten as

$$\hat{H}_n = E_n(\boldsymbol{R}^0) + \sum_\alpha \left[-\frac{\hbar^2}{2}\frac{\partial^2}{\partial Q_\alpha^2} + \frac{1}{2}\omega_\alpha^2 \left(Q_\alpha - \frac{|V_\alpha|}{\omega_\alpha^2} \right)^2 - \frac{V_\alpha^2}{2\omega_\alpha^2} \right]. \tag{2.82}$$

It should be noted that the $|V_\alpha|$ is equal to the force f_α along Q_α. Therefore, the molecular deformation from $\boldsymbol{R}^0$ by $(|V_\alpha|/\omega_\alpha^2)$ occurs because of the linear vibronic coupling constant V_α. The molecular deformation yields the stabilization energy, or reorganization energy (Fig. 2.11)

$$\Delta E = \sum_\alpha \frac{V_\alpha^2}{2\omega_\alpha^2}. \tag{2.83}$$

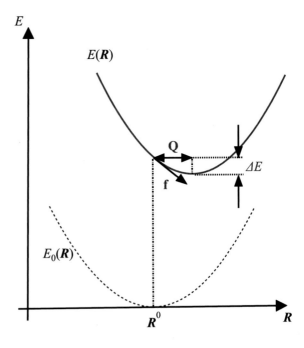

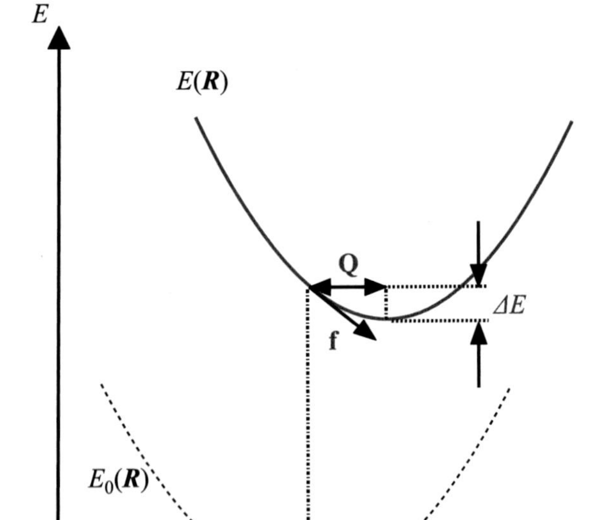

Fig. 2.11 Force acting on a molecule: linear vibronic couplings

2.7.2 Spontaneous Symmetry Breaking of Non-linear Molecule: Jahn–Teller Effect

For degenerate electronic states of a non-linear molecule A_3 with energy E_n, for example, $\Psi_{E\theta}$ and $\Psi_{E\epsilon}$, Eq. (2.79) becomes

$$
\begin{aligned}
(\hat{\mathbf{H}})_{mn} &= E_n(\boldsymbol{R}^0)\delta_{mn} + \sum_{\alpha} -\frac{\hbar^2}{2}\frac{\partial^2}{\partial Q_\alpha^2}\delta_{mn} + V_{mn,\alpha}Q_\alpha \\
&\quad + \sum_{\alpha,\beta}\frac{1}{2}W_{mn,\alpha\beta}Q_\alpha Q_\beta + \cdots \quad (m, n = \theta, \epsilon).
\end{aligned}
\tag{2.84}
$$

Since the symmetric product is

$$
[E^2] = A_1 + E,
\tag{2.85}
$$

and non-zero linear vibronic couplings arise for E and A_1 vibrational modes. Among these modes, E modes break the molecular symmetry. When we consider the vibrational coordinates (Q_θ, Q_ϵ), we obtain the following Jahn–Teller Hamiltonian:

$$
\begin{aligned}
(\hat{\mathbf{H}})_{mn} &= E_n(\boldsymbol{R}^0)\delta_{mn} + \sum_{\alpha=\theta,\epsilon} -\frac{\hbar^2}{2}\frac{\partial^2}{\partial Q_\alpha^2}\delta_{mn} + V_{mn,\alpha}Q_\alpha \\
+ \sum_{\alpha,\beta=\theta,\epsilon} &\frac{1}{2}W_{mn,E}{Q_\alpha Q_\beta}_E + \frac{1}{2}W_{mn,A_1}{Q_\alpha Q_\beta}_{A_1} + \cdots \quad (m, n = \theta, \epsilon).
\end{aligned}
\tag{2.86}
$$

We assume C_{3v} symmetry. From the Wigner–Eckart theorem and Clebsch–Gordan coefficients,

$$
\hat{\mathbf{H}} = -\frac{\hbar^2}{2}\left(\frac{\partial^2}{\partial Q_\theta^2} + \frac{\partial^2}{\partial Q_\epsilon^2}\right)\boldsymbol{\sigma}_0 + \hat{\mathbf{U}}(Q_\theta, Q_\epsilon),
\tag{2.87}
$$

where the potential energy matrix $\hat{\mathbf{U}}(Q_\theta, Q_\epsilon)$ is defined by

$$
\begin{aligned}
\hat{\mathbf{U}}(Q_\theta, Q_\epsilon) &= E_n(\boldsymbol{R}^0)\boldsymbol{\sigma}_0 + V_E\left(Q_\theta\boldsymbol{\sigma}_z + Q_\epsilon\boldsymbol{\sigma}_x\right) + \frac{1}{2}W_E\left[\left(Q_\theta^2 - Q_\epsilon^2\right)\boldsymbol{\sigma}_x - \left(2Q_\theta Q_\epsilon\right)\boldsymbol{\sigma}_y\right] \\
&\quad + \frac{1}{2}W_{A_1}\left(Q_\theta^2 + Q_\epsilon^2\right)\boldsymbol{\sigma}_0,
\end{aligned}
\tag{2.88}
$$

and Pauli's spin matrices are defined by

$$
\boldsymbol{\sigma}_0 = \begin{pmatrix} 1 & 0 \\ 0 & 1 \end{pmatrix}, \quad \boldsymbol{\sigma}_x = \begin{pmatrix} 0 & 1 \\ 1 & 0 \end{pmatrix}, \quad \boldsymbol{\sigma}_y = \begin{pmatrix} 0 & -i \\ i & 0 \end{pmatrix}, \quad \boldsymbol{\sigma}_z = \begin{pmatrix} 1 & 0 \\ 0 & -1 \end{pmatrix}.
\tag{2.89}
$$

Polar coordinates are introduced as

$$\begin{cases} Q_\theta = \rho \cos\phi \\ Q_\epsilon = \rho \sin\phi \end{cases}. \tag{2.90}$$

The diagonalization of the potential energy matrix (2.88) yields

$$E_\pm(\rho,\phi) = E_n(\boldsymbol{R}^0) + \frac{1}{2}W_{A_1}\rho^2 \pm \sqrt{V_E^2\rho^2 + V_E W_E \rho^3 \cos 3\phi + W_E^2\rho^4}. \tag{2.91}$$

Figure 2.12 shows the lower energy $E_-(\rho,\phi)$. There exist three minimum positions which are equivalent structures.

2.7.3 *Spontaneous Symmetry Breaking of Non-linear Molecule: Pseudo-Jahn–Teller Effect*

In the previous subsections, we discuss the cases in which the electronic state(s) with the energy E_n is considered. We discuss the non-degenerate state Ψ_n considering other state Ψ_m with a different energy E_m ($E_m \neq E_n$, $E_n < E_m$). For simplicity, we take two energy levels, $n = 1$ and $m = 2$. State Ψ_m can be degenerate; however, we assume Ψ_m also non-degenerate here. Since Ψ_n and Ψ_m are non-degenerate, the irreducible representations $\Gamma_{n/m}$ are one dimensional. The vibrational modes that couple with Ψ_n and Ψ_m are non-degenerate with $\Gamma = \Gamma_n \times \Gamma_m$. We take a single mode Q with Γ here. The vibronic Hamiltonian (2.76) becomes

$$\hat{\mathbf{H}}_{\mathrm{PJT}} = -\frac{\hbar^2}{2}\frac{\partial^2}{\partial Q^2}\boldsymbol{\sigma}_0 + \hat{\mathbf{U}}_{\mathrm{PJT}}(Q), \tag{2.92}$$

where the potential energy matrix is defined by

$$\hat{\mathbf{U}}_{\mathrm{PJT}}(Q) = \begin{pmatrix} E_1(\boldsymbol{R}^0) & 0 \\ 0 & E_2(\boldsymbol{R}^0) \end{pmatrix} + V_{12} Q \boldsymbol{\sigma}_x + \frac{1}{2}\begin{pmatrix} W_{11} & 0 \\ 0 & W_{22} \end{pmatrix} Q^2. \tag{2.93}$$

When the second term in Eq. (2.93) can be regarded as a perturbation, the lower eigenenergy of Eq. (2.93) is, up to the second order,

$$E_1(Q) = E_1(\boldsymbol{R}^0) + \frac{1}{2}W_{\mathrm{PJT}}Q^2 + \cdots, \tag{2.94}$$

where

$$W_{\mathrm{PJT}} = W_{11} - \frac{|V_{12}|^2}{E_2(\boldsymbol{R}^0) - E_1(\boldsymbol{R}^0)}. \tag{2.95}$$

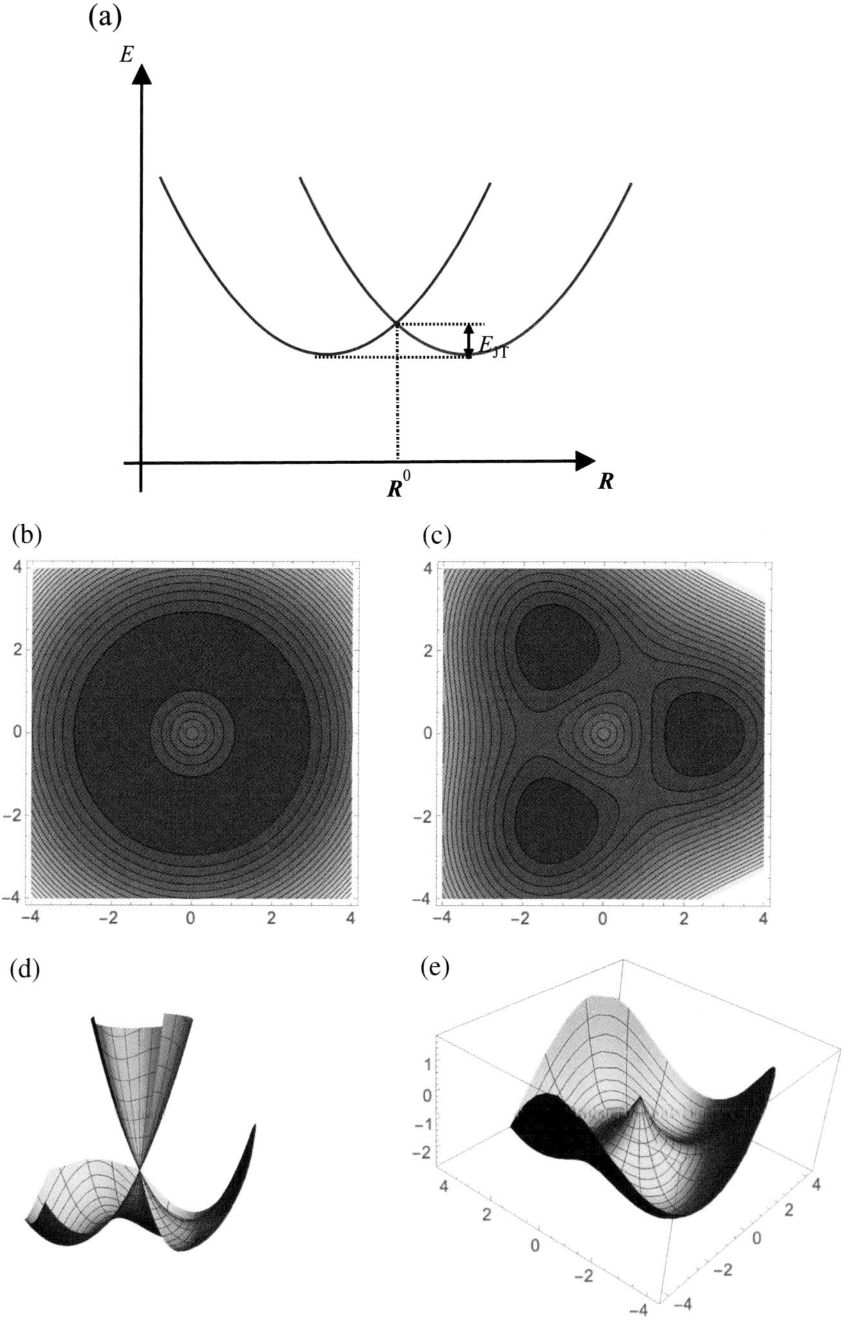

Fig. 2.12 Spontaneous symmetry breaking of a non-linear molecule: Jahn–Teller effect. **a**, **b** JT potential energy surface for $W_E = 0$. Note that the energy minima are infinitely degenerate (trough). **c**–**e** JT potential energy surface for $W_E \neq 0$. There appears three minima and three maxima in the trough

The first term in Eq. (2.95) is positive. If the off-diagonal vibronic coupling constant V_{12} is large, or the energy difference between $E_1(\boldsymbol{R}^0)$ and $E_2(\boldsymbol{R}^0)$ is small, W_{PJT} can be negative. For such a case, the reference structure $\boldsymbol{R}^0$ is not stable. Therefore, if Γ_2 is different from Γ_1, the irreducible representation of the vibrational mode is not totally symmetric, leading to a deformed structure with lower symmetry.

2.7.4 Spontaneous Symmetry Breaking of Linear Molecule: Renner–Teller Effect

In Sect. 2.7.2, we excluded linear molecules from the discussion. For a linear molecule with degenerate electronic states, the linear vibronic coupling term is vanishing.

Here we assume degenerate electronic states of a molecule with $D_{\infty h}$ symmetry. The ground state of linear NH_2 discussed in Sect. 1.2 is such an electronic state. Since the lowest order of the non-vanishing vibronic couplings is quadratic ones (see Sect. 5.2.3), instead of Eq. (2.86), vibronic Hamiltonian can be written as

$$
\begin{aligned}
(\hat{\mathbf{H}})_{mn} &= E_n(\boldsymbol{R}^0)\delta_{mn} + \sum_{\alpha=\theta,\epsilon} -\frac{\hbar^2}{2}\frac{\partial^2}{\partial Q_\alpha^2}\delta_{mn} + \\
&+ \sum_{\alpha,\beta=\theta,\epsilon} \frac{1}{2}W_{mn,E}\{Q_\alpha Q_\beta\}_E + \frac{1}{2}W_{mn,A_1}\{Q_\alpha Q_\beta\}_{A_1} + \cdots , \qquad (2.96)
\end{aligned}
$$

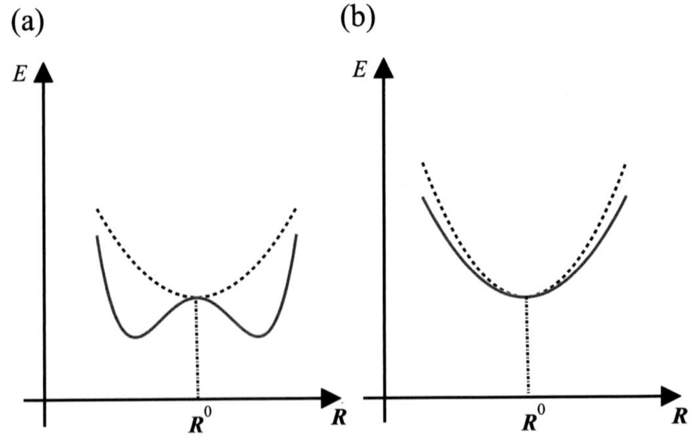

Fig. 2.13 Spontaneous symmetry breaking of a non-linear molecule: pseudo-Jahn–Teller effect

where Q_α and Q_β denote the coordinates of the E_u mode, which are bending modes shown in Fig. 5.1. The quadratic vibronic coupling terms in (2.96) give rise to lowering the vibrational frequency of the degenerate mode or a bent molecular structure. This effect is called the Renner–Teller effect (Fig. 2.13).

Reference

1. Ceulemans AJ (2013) Group theory applied to chemistry. Springer, Dordrecht

Chapter 3
Vibronic Coupling Density Analyses for Molecular Deformation

Abstract The basis of the vibronic coupling density (VCD) analysis has been introduced in Chap. 2. The vibronic coupling constant (VCC) as well as the VCD will be exactly defined in Sect. 5.4.1. As seen in Sect. 5.4.1, both are starting from the *ab initio* molecular Hamiltonian, and systematic, rational ways to understand chemical phenomena, and which can give the quantitative evaluation of the force applied under the chemical deformation process. We offer the approach of chemistry with more rational and general way through the visualization by VCD and the evaluation by VCC. Vibronic coupling is the interaction between vibrational and electronic motions. The local picture of a vibronic coupling can be expressed in terms of electronic and vibrational structures using VCD. We describe the concepts of VCC and VCD within the crude adiabatic approximation.

Keywords Vibronic coupling density · Vibronic coupling constant · Elongation · Orbital relaxation · Renner–Teller effect · Pseudo-Jahn–Teller effect · Orbital vibronic coupling density · Orbital overlap · Potential derivative · Imaginary mode · Quadratic vibronic coupling density · Jahn–Teller effect · Effective vibronic coupling density · Regioselectivity · Structural relaxation

3.1 Elongation Force on Anionization of H_2 and C_2H_4

The physical meaning of the diagonal element of vibronic coupling is a force. A change in the electronic state causes a force to act between nuclei and gives rise to a geometry change. Therefore, we can understand the structural change by analyzing the vibronic coupling density (VCD). As defined in Sect. 5.4.1, a linear vibronic coupling constant (VCC), $V_{mn,\alpha}$, is described as an integral of a VCD, $\eta_{mn,\alpha}(\boldsymbol{x})$ [1–4],

$$V_{mn,\alpha} = \int \eta_{mn,\alpha}(\boldsymbol{x}) d\boldsymbol{x}, \tag{3.1}$$

$$\eta_{mn,\alpha}(\boldsymbol{x}) := \begin{cases} \Delta\rho_m(\boldsymbol{x}) \times v_\alpha(\boldsymbol{x}) \ (m = n), \\ \rho_{mn}(\boldsymbol{x}) \times v_\alpha(\boldsymbol{x}) \ \ (m \neq n). \end{cases} \tag{3.2}$$

T. Kato et al., *Vibronic Coupling Density*,
SpringerBriefs in Molecular Science,
https://doi.org/10.1007/978-981-16-1796-6_3

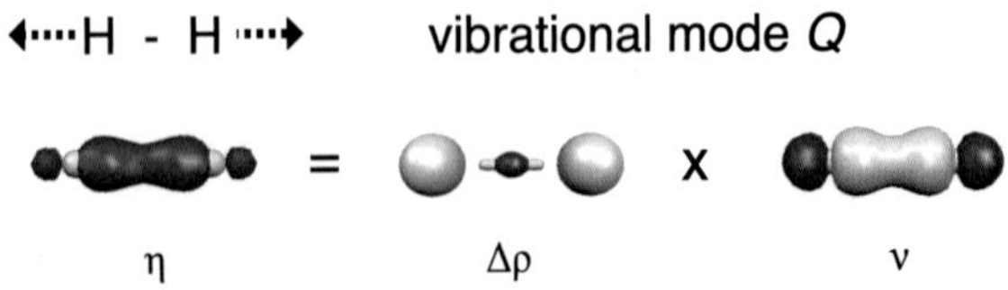

Fig. 3.1 Vibronic coupling density analysis for hydrogen molecule anion (ROHF/6-31G with first derivatives). Top: vibrational mode, lower: vibronic coupling density η, electron-density difference $\Delta\rho \times$ potential derivative v. The blue and gray surfaces denote negative and positive densities, respectively. Reprinted by permission from Springer Nature, The Jahn–Teller Effect: Fundamentals and Implications for Physics and Chemistry, Springer Series in Chemical Physics, vol. 97, ed. by H. Köppel, D. R. Yarkony, H. Barentzen (Springer-Verlag, Berlin, 2009)

The concept of VCD enables us to visually and intuitively analyze the origin of vibronic coupling in terms of $\Delta\rho_m(\boldsymbol{x})$ or $\rho_{mn}(\boldsymbol{x})$, which is an electron-density difference or an overlap electron density obtained from electronic structures, and $v_\alpha(\boldsymbol{x})$, which is an electron potential energy derivation obtained from vibrational structures. The VCD can be generally understood as the three-dimensional picture, which describes the interaction between molecular vibration and electron density.

As an example, the vibronic coupling density analysis for a hydrogen molecule anion is shown in Fig. 3.1. As mentioned in Sect. 1.1, when a neutral hydrogen molecule acquires an electron, and the chemical bond will be elongated since the additional electron occupies the anti-bonding LUMO. The vibronic coupling density analysis reveals this driving force [3].

From Fig. 3.1, it is seen that most of the negative vibronic coupling density, the blue surface occurs in the bond region, that is, the negative gradient of a potential energy surface with respect to the vibration of nuclei moving apart. This is the driving force of the chemical bond being elongated on anionization. The vibronic coupling density is a product between the electron-density difference $\Delta\rho_i$ and the potential derivative v_α, as defined in Sect. 5.4.1. More specifically, v_α is the derivative of the potential acting on a single electron from all nuclei for the vibrational coordinate of mode α. In the case of the hydrogen molecule,

$$
\begin{aligned}
v_\alpha(\boldsymbol{x}) &\sim \frac{\partial}{\partial(-X_{\mathrm{H_A}})}\left(-\frac{1}{|\boldsymbol{x}-\boldsymbol{R}_{\mathrm{H_A}}|}\right)+\frac{\partial}{\partial X_{\mathrm{H_B}}}\left(-\frac{1}{|\boldsymbol{x}-\boldsymbol{R}_{\mathrm{H_B}}|}\right)\\
&=\frac{x-X_{\mathrm{H_A}}}{|\boldsymbol{x}-\boldsymbol{R}_{\mathrm{H_A}}|^3}-\frac{x-X_{\mathrm{H_B}}}{|\boldsymbol{x}-\boldsymbol{R}_{\mathrm{H_B}}|^3},
\end{aligned}
\tag{3.3}
$$

where its molecular axis is taken as the x-axis. Thus, $v_\alpha(\boldsymbol{x})$ has p_x-type distribution ($p_x \sim x/r^3$) around each nucleus because it is the linear derivative of the totally symmetric potentials around the nuclei. Negative distribution appears in the positive direction of the vibrational mode. As with the present case, in general, a potential derivative always has p-type distribution in the corresponding vibrational direction. In Fig. 3.1, it is noted that small negative $\Delta\rho_i$ occurs in the bond region. Because the additional electron distribution represented by the gray surfaces in Fig. 3.1 polarizes

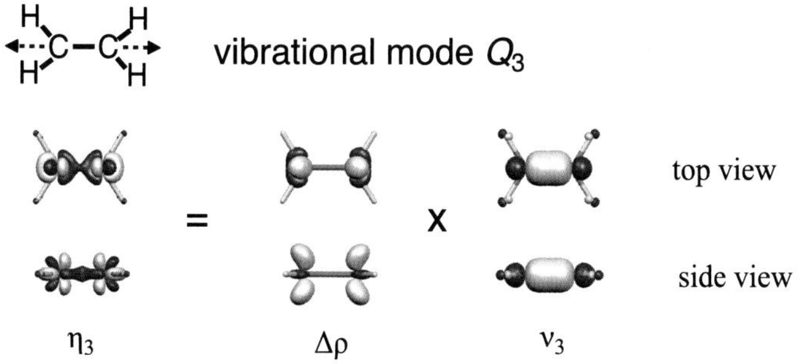

Fig. 3.2 Top: Vibrational mode, lower: top view and side view of vibronic coupling density η_3, electron-density difference $\Delta\rho \times$ potential derivative v_3. The blue and gray surfaces denote negative and positive densities, respectively. Reprinted by permission from Springer Nature, The Jahn–Teller Effect: Fundamentals and Implications for Physics and Chemistry, Springer Series in Chemical Physics, vol. 97, ed. by H. Köppel, D. R. Yarkony, H. Barentzen (Springer–Verlag, Berlin, 2009)

orbitals occupied by the other electrons. The negative $\Delta\rho_i$ couples with the positive potential derivative v_α.[1] The orbital polarization due to anionization or orbital relaxation plays a crucial role in vibronic coupling.

The importance of orbital relaxation can be observed in π electron systems. Figure 3.2 shows the vibronic coupling density analysis for the ethylene anion. The carbon–carbon stretching mode among 12 vibrational modes is focused in terms of the vibronic coupling via the anionization process. Anionization of ethylene gives rise to an elongation of the double bond. As shown in Fig. 3.2, even the additional electron occupies the anti-bonding π LUMO, a negative vibronic coupling density $\eta_{m,\alpha}$ occurs near the carbon atoms in the molecular plane. The negative $\eta_{m,\alpha}$ is originated from the negative electron-density difference $\Delta\rho_m$ due to the electron polarization of additional π electron.

These simple examples clearly show that orbital relaxation is crucial in vibronic coupling. Therefore, variationally optimized wave functions should be employed for vibronic coupling calculations. The frozen orbital approximation is not suitable for calculation [2].

3.2 Molecular Deformation of NH_2 and H_2O by Renner–Teller Effect and Pseudo-Jahn–Teller Effect

In the initial nuclear configuration $\boldsymbol{R}_0$ of the linear structure, the point group is $D_{\infty h}$, the normal modes shown in Fig. 3.3 can be defined. The energy of the molecule including the vibrational deformation described by normal coordinates $\boldsymbol{Q}$ is given by the eigenenergies of the molecular Hamiltonian,

[1] The positive value of the potential derivative v_α between two proton nuclei is due to the increase of the electronic potential with separating the bonding electrons and two proton nuclei apart.

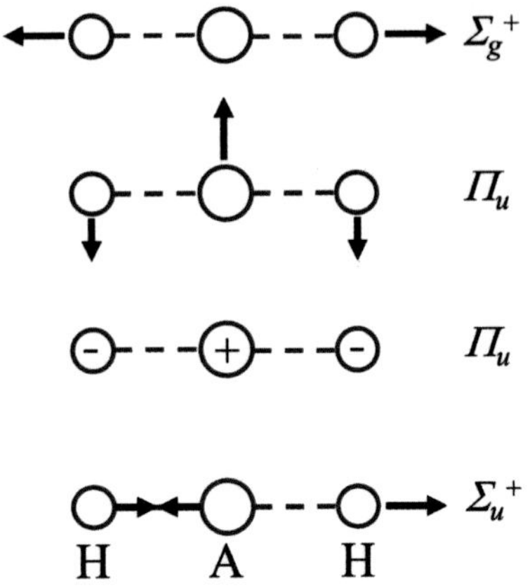

Fig. 3.3 Vibrational normal modes of $H_2A(D_{\infty h})$

$$H = H_0 + \left(\frac{\partial U}{\partial Q}\right)_0 Q + \frac{1}{2}\left(\frac{\partial^2 U}{\partial Q^2}\right)_0 Q^2 + \cdots, \tag{3.4}$$

where H_0 is the Hamiltonian at $\boldsymbol{R}_0$, U is the interaction potential energy between nuclei and electrons and among nuclei. Using the perturbation method, the lowest eigenenergy E_0 is obtained as

$$E = E_0 + \langle\Psi_0|\left(\frac{\partial U}{\partial Q}\right)_0|\Psi_0\rangle Q + \frac{1}{2}\langle\Psi_0|\left(\frac{\partial^2 U}{\partial Q^2}\right)_0|\Psi_0\rangle Q^2 + \sum_{k=1}^{\infty}\frac{|\langle\Psi_0|\left(\frac{\partial U}{\partial Q}\right)_0|\Psi_k\rangle|^2 Q^2}{E_0 - E_k}, \tag{3.5}$$

as mentioned in Sects. 2.7.3 and 2.7.4 . The third term causes the degenerate energy split with the deformation, which is named the Renner–Teller effect [5]. The integral $\langle\Psi_0|\left(\frac{\partial^2 U}{\partial Q^2}\right)_0|\Psi_0\rangle$ is not zero, and this term is represented as the matrix form for the degenerate Ψ_0.

In the linear structure of $NH_2(D_{\infty h})$, nine electrons of NH_2 configured on lower lying five MOs depicted in Fig. 3.4, resulting in the doubly degenerate Π_u ground state, and the third term in Eq. 3.5 becomes the two by two matrix,

$$\langle\Psi_0|\left(\frac{\partial^2 U}{\partial Q^2}\right)_0|\Psi_0\rangle = \begin{pmatrix} \langle\Psi_{01}|\left(\frac{\partial^2 U}{\partial Q^2}\right)_0|\Psi_{01}\rangle & \langle\Psi_{01}|\left(\frac{\partial^2 U}{\partial Q^2}\right)_0|\Psi_{02}\rangle \\ \langle\Psi_{02}|\left(\frac{\partial^2 U}{\partial Q^2}\right)_0|\Psi_{01}\rangle & \langle\Psi_{02}|\left(\frac{\partial^2 U}{\partial Q^2}\right)_0|\Psi_{02}\rangle \end{pmatrix}, \tag{3.6}$$

among four matrix elements, non-zero integral $\langle\Psi_i|\left(\frac{\partial^2 U}{\partial Q^2}\right)_0|\Psi_j\rangle \neq 0, (i, j = 01, 02)$ is obtained against the vibrational mode of Σ_g^+ and Π_u, because of $\Pi_u \times \Pi_u = \Sigma_g^+ + \Sigma_u^+ + \Delta_g$ as shown in Table 5.9 and that the representation of $\left(\frac{\partial^2 U}{\partial Q^2}\right)_0$ is identical with that of Q^2. However, the mode Σ_g^+ or Σ_u^+ of the deformation $\boldsymbol{Q}$ does not contribute to breaking the axial symmetry of the molecule, as shown in Fig. 3.3. On the other hand, the Π_u nuclear displacement demolishes the axial symmetry. As a result, the non-zero cross-terms of the matrix $\langle\Psi_{01}|\left(\frac{\partial^2 U}{\partial Q^2}\right)_0|\Psi_{02}\rangle$ and $\langle\Psi_{02}|\left(\frac{\partial^2 U}{\partial Q^2}\right)_0|\Psi_{01}\rangle$ give rise to the energy split between the doubly degenerate Π_u state concerning the deformation out of axis Π_u.

In the case of H_2O, the fourth term in Eq. 3.5 causes energy stabilization with the deformation by the pseudo-Jahn–Teller effect [6]. The denominator E_0–E_k is minus because E_k is the upper energy level of the excited state. If the integral $\langle\Psi_0|\left(\frac{\partial U}{\partial Q}\right)_0|\Psi_k\rangle$ is not zero. This term always gives a minus energy corresponding to the stabilization with deformation. In the linear structure of $D_{\infty h}$, the electronic configuration of the H_2O ground state is shown in Fig. 3.4. The lowest excited state corresponds to the excitation $\pi_u \longrightarrow 3\sigma_g$. When the molecule is deformed along with the normal mode of Π_u, which corresponds to the bent deformation as shown in Fig. 3.3, the value of the integral $\langle\Psi_0|\left(\frac{\partial U}{\partial Q}\right)_0|\Psi_k\rangle$ will be non-zero because $\pi_u \times \Pi_u \times \sigma_g$ includes the totally symmetric representation of Σ_g (see Table 5.9). The VCD analysis can visually understand this. Within the orbital approximation, $\langle\Psi_0|\left(\frac{\partial U}{\partial Q}\right)_0|\Psi_k\rangle$ is given by the integral of the OVCD that is the product of the MO overlap density $\rho_{\mathrm{HO\,LU}}$ between the $\pi_u(\xi)$ HOMO and the σ_g LUMO with the potential derivative $v_{\Pi_u(\xi)}$ for the $\Pi_u(\xi)$ mode ($\xi = x$ or y). Note that the π_u HOMOs are doubly degenerate and distinguished by x and y. The Π_u vibrational modes are also such a case. Figure 3.5 shows that $\rho_{\mathrm{HO\,LU}}$ and v_{Π_u} have the same pattern of distributions, $(-, +, -)$ on the H–O–H linear structure, and hence their product, i.e., the OVCD, has the same sign distribution over the molecule. Accordingly, the integral of the OVCD has a large non-zero value, leading to the pseudo-Jahn–Teller

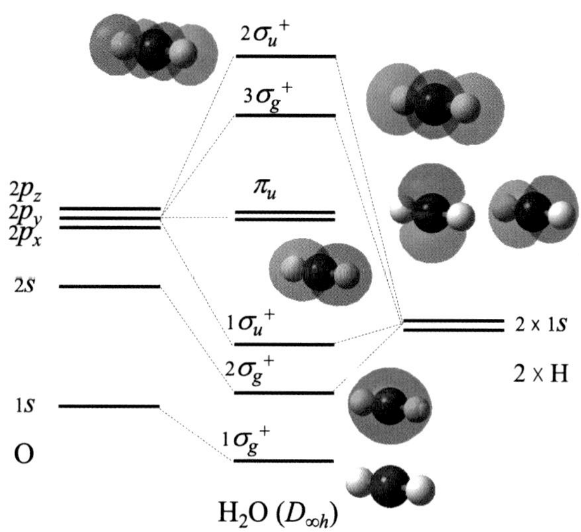

Fig. 3.4 Molecular orbitals of water molecule in linear structure of $D_{\infty h}$ symmetry

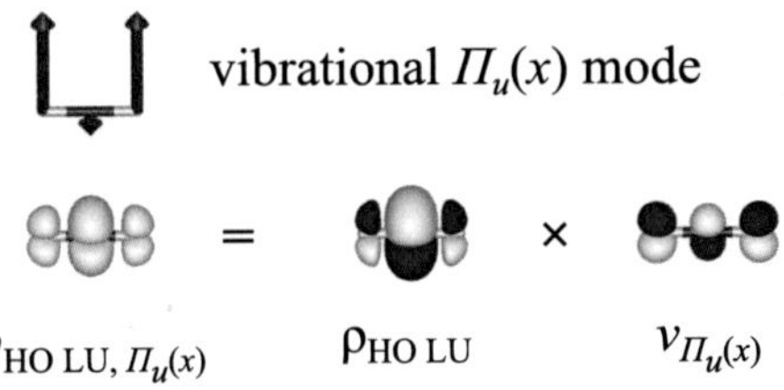

Fig. 3.5 Upper: imaginary $\Pi_u(x)$ mode, lower: orbital vibronic coupling density (OVCD) = orbital overlap density $\rho_{\mathrm{HO\,LU}}$ × potential derivative $v_{\Pi_u(x)}$. The blue and gray surfaces denote negative and positive densities, respectively

effect due to the mixing of the excited state with the ground one via the vibronic coupling $\langle\Psi_0|\left(\frac{\partial U}{\partial Q}\right)_0|\Psi_k\rangle Q$. Consequently, H_2O molecule is stabilized along the bent deformation.

3.3 Vibronic Coupling Density of Ammonia Molecule NH_3

The molecular structure of NH_3 can also be well explained by the pseudo-Jahn–Teller effect [6]. The vibrational modes of $NH_3(D_{3h})$ are given, as shown in Fig. 3.6

$$\Gamma_{vib} \quad = \quad A_1' \quad + \quad A_2'' \quad + \quad 2E'. \tag{3.7}$$

From the vibrational analysis by the calculation at the B3LYP/STO-3G level of theory, we obtained an A_2'' mode with an imaginary frequency, $902.3i$ cm^{-1}.[2] The molecule would be deformed and stabilized along the direction of the imaginary A"$_2$ mode. The pseudo-Jahn–Teller energy curve of the planar D_{3h} NH_3 model is given by

$$E = E_0 + \frac{1}{2}\langle\Psi_0|\left(\frac{\partial^2 U}{\partial Q^2}\right)_0|\Psi_0\rangle Q^2 + \frac{|\langle\psi_{a_2''}|\left(\frac{\partial U}{\partial Q}\right)_0|\psi_{a_1'}\rangle|^2 Q^2}{E_{a_2''} - E_{a_1'}}, \tag{3.8}$$

as mentioned in Sect. 2.1, where Ψ_0 denotes the ground electronic wave function, $\psi_{a_2''}$ is the HOMO, $\psi_{a_1'}$ is the LUMO, $E_{a_2''}$ is the energy level of the HOMO, and $E_{a_1'}$ is the energy level of the LUMO. The third term on the right-hand side corresponds to the lowest unoccupied orbital mixing of $a_2'' \longrightarrow a_1'$. When the orbital mixing is led via the A_2'' mode, the integral $\langle\psi_{a_2''}|\left(\frac{\partial U}{\partial Q}\right)_0|\psi_{a_1'}\rangle$ is non-zero because the product of $a_2'' \times A_2'' \times a_1'$ includes the totally symmetric representation of A_1' as referred

[2] The imaginary mode means the vibrational mode having an imaginary frequency, $\omega_\alpha = \frac{1}{2\pi c_0}\sqrt{\frac{\kappa_\alpha}{m}}$ [cm^{-1}]. The harmonic oscillator with an imaginary ω_α has a minus force constant κ_α, which corresponds to not an oscillation on a concave surface but a motion on a convex one. The movement on a convex surface is the rolling down along the coordinate of the mode.

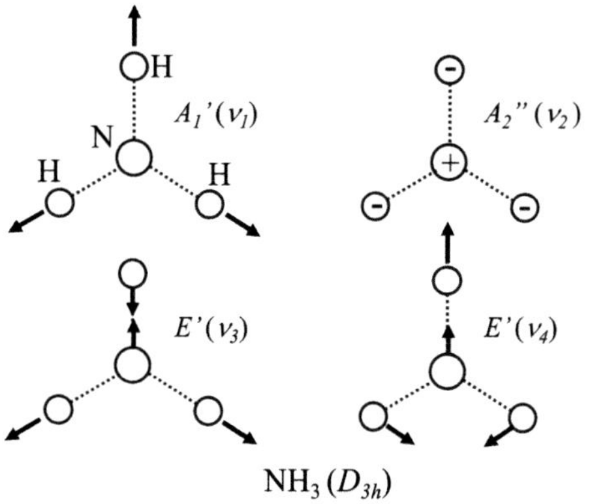

Fig. 3.6 Vibrational normal modes of NH_3 assuming the D_{3h} structure

in Table 5.10, and as a result, the third term gives the energy stabilization in the direction of the A_2'' mode. However, the actual stabilization is determined by the balance between the second and third terms. Taking the VCD analysis into account, we can estimate the ratio of the magnitude between the second and third terms as described below.

If the pseudo-Jahn–Teller effect is discussed by using vibronic coupling density (VCD) scheme, it is much advantageous that we can see the effect in a visualized manner [4]. The VCD scheme gives that the orbital vibronic coupling density (OVCD) is localized on the nitrogen atoms, and the instability force originated from the vibronic coupling of the orbital overlap between the lone pair of the nitrogen atom via the umbrella bending mode.

The second term is given by the quadratic vibronic coupling density (QVCD), as defined in Sect. 5.4.2,

$$\eta_{A_1', A_2'' A_2''} := \rho_{A_1'} \times w_{A_2'' A_2''}, \tag{3.9}$$

and the third term is given by the OVCD, as described in Sect. 5.4.4.2.

$$\eta_{HO\,LU, A_2''} := \rho_{HO\,LU} \times v_{A_2''}. \tag{3.10}$$

The QVCD is shown in Fig. 3.7. The electron density ρ is totally symmetric, and mainly distributed on the nitrogen atom. However, since the quadratic potential derivative $\omega_{a_2''}$ has small distribution on the nitrogen atom, and the QVCD is almost equally distributed with blue- and gray-colored surfaces on the nitrogen and hydrogen atoms. The integral of the QVCD on each atom becomes small with the cancellation between blue (negative) and gray (positive) densities. On the other hand, the OVCD shown in Fig. 3.8 is localized on the nitrogen atoms with the only blue-colored surface, which means the pseudo-Jahn–Teller instability originates from the lone pair of the nitrogen atoms.

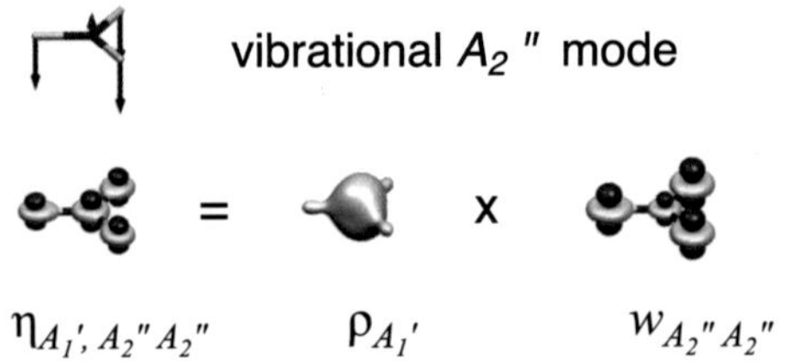

Fig. 3.7 Upper: imaginary A_2'' mode, lower: quadratic vibronic coupling density ($\eta_{A_1',A_2''A_2''}$) = electron density $\rho_{A_1'}$ × quadratic potential derivative $w_{A_2''A_2''}$. The blue and gray surfaces denote negative and positive densities, respectively. Reprinted from J. Phys.: Conf. Ser., vol. 428, T. Sato, M. Uejima, N. Iwahara, N. Haruta, K. Shizu, and K. Tanaka, Vibronic coupling density and related concepts, 012010 (2013)

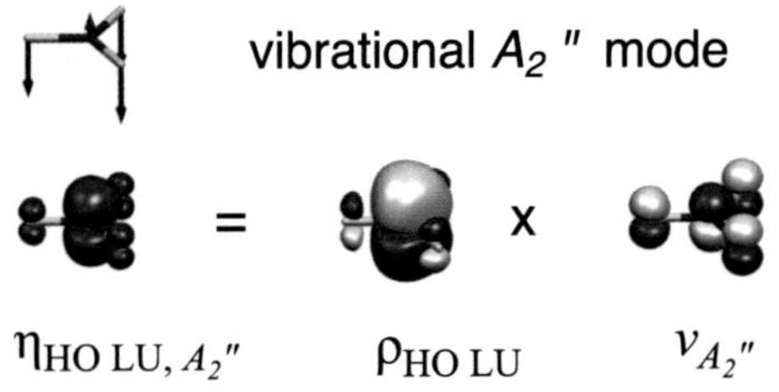

Fig. 3.8 Upper: imaginary A_2'' mode, lower: orbital vibronic coupling density ($\eta_{\mathrm{HO\ LU},A_2''}$) = orbital overlap density $\rho_{\mathrm{HO\ LU}}$ × potential derivative $v_{A_2''}$. The blue and gray surfaces denote negative and positive densities, respectively. Reprinted from J. Phys.: Conf. Ser., vol. 428, T. Sato, M. Uejima, N. Iwahara, N. Haruta, K. Shizu, and K. Tanaka, Vibronic coupling density and related concepts, 012010 (2013)

3.4 Vibronic Coupling Density Explanation for Zigzag Conformation of 6-Cycloparaphenylene

Here we introduce the interpretation by the pseudo-Jahn–Teller effect (PJTE) to interpret the tautomerism due to the ionization of 6-CPP molecule. So we visualize the origin of the vibronic interaction using VCD for the deformation to the zigzag conformation from belt-like one of 6-CPP. The MO calculation using the RB3LYP/STO-3G level was done for optimizing the geometry of 6-CPP within the D_{6h} symmetry of the belt-like conformation [7]. The vibrational analysis gives the imaginary frequency of the B_{1g} normal mode shown in Fig. 3.9. The imaginary frequency indicates that the D_{6h} structure is unstable for the displacement along this B_{1g} normal mode, and the symmetry lowers into D_{3d} with non-zero dihedral angles between the neighboring benzene units.

The electronic ground state is calculated as shown in Fig. 3.10, in which the HOMO orbital is π-orbital with a_{2g} symmetry. As mentioned in Sect. 1.3, $\{\langle\psi_{a_2''}|\left(\frac{\partial U}{\partial Q}\right)_0|\psi_{a_1'}\rangle Q\}^2/(E_{a_2''} - E_{a_1'})$ in Eq. 3.8 gives the energy stabilization with the displacement along Q. When the displacement is along the B_{1g} normal mode,

Fig. 3.9 The B_{1g} vibrational normal mode of 6-CPP in D_{6h} symmetry with the imaginary frequency. Reprinted from Chem. Phys. Lett., vol. 598, Y. Kameoka, T. Sato, T. Koyama, K. Tanaka, and T. Kato, Pseudo-Jahn–Teller origin of distortion in 6-cycloparaphenylene, 69 (2014), with permission from Elsevier

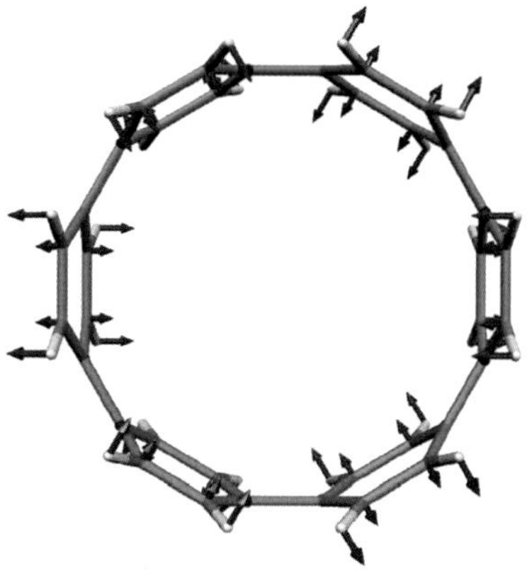

Fig. 3.10 Molecular orbitals obtained by the MO calculation. The HOMO for the neutral ground state of 6-CPP is a_{2g}, and the next HOMOs are e_{1u} with a degeneracy. The b_{2g}-orbital is an unoccupied level which leads to the pseudo-JT effect. Reprinted from Chem. Phys. Lett., vol. 598, Y. Kameoka, T. Sato, T. Koyama, K. Tanaka, and T. Kato, Pseudo-Jahn–Teller origin of distortion in 6-cycloparaphenylene, 69 (2014), with permission from Elsevier

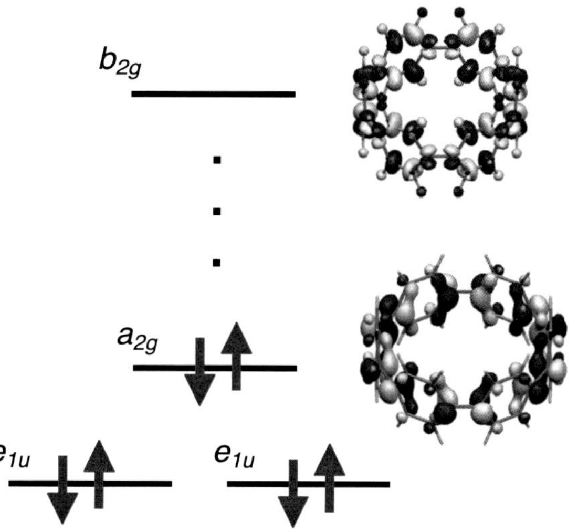

the term of the pair with $\psi_{a_2''} = a_{2g}$(HOMO) and $\psi_{a_1'} = b_{2g}$ (unoccupied orbital) yields a maximum coupling because of the selection rule: $a_{2g} \times B_{1g} \times b_{2g} = a_{1g}$. And this pair of orbitals make the largest contribution to the instability of the belt-like structure, that is, the orbital pair including the HOMO plays an important role in the structural change to the zigzag conformation.

The OVCD analysis of the pair of the HOMO and the b_{2g} unoccupied orbital is shown in Fig. 3.11. The symmetries and phase patterns in the pair of orbitals play an crucial role in OVCC. The b_{2g} unoccupied orbital is a σ-orbital, while the HOMO is a π-orbital, as shown in Fig. 3.11, respectively. Therefore, the overlap density $\rho_{a_{2g}\,b_{2g}}$ between them has π-type distribution. The sign of $\rho_{a_{2g}\,b_{2g}}$ is opposite on both sides of the benzenes of 6-CPP. The function $v_{B_{1g}}$ also has a π-type distribution

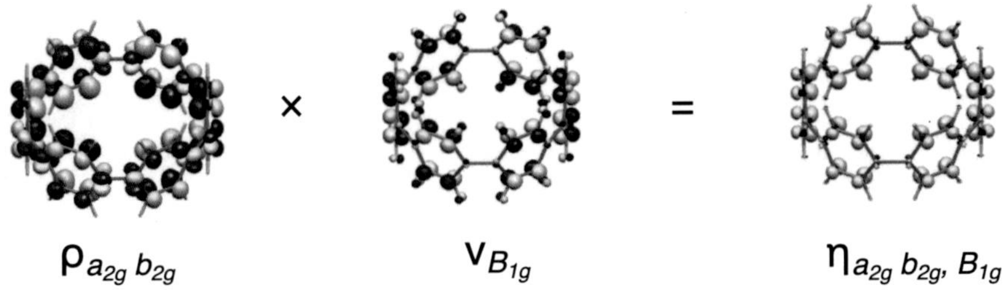

Fig. 3.11 VCD of 6-CPP is obtained by the product of the overlap integral with the potential derivative along the coordinate of the imaginary B_{1g} vibrational normal mode. Reprinted from Chem. Phys. Lett., vol. 598, Y. Kameoka, T. Sato, T. Koyama, K. Tanaka, and T. Kato, Pseudo-Jahn–Teller origin of distortion in 6-cycloparaphenylene, 69 (2014), with permission from Elsevier

because of the displacement of the B_{1g} mode. In addition, the signs of $\nu_{B_{1g}}$ and $\rho_{a_{2g}\,b_{2g}}$ coincide almost everywhere. Accordingly, OVCD $\eta_{a_{2g}\,b_{2g},B_{1g}}$ has a positive value almost everywhere. The orbital phases and symmetries of the orbital pairs give rise to a large value of the OVCD with the same sign.

Since the vibronic coupling between the orbital pair, including the HOMO plays the crucial role in the structural change to the zigzag conformation, the ionization from the HOMO orbital gives rise to the increasing contribution of the quinoid form in the cation and dication species of 6-CPP [8].

3.5 Vibronic Coupling Density and Jahn–Teller Hamiltonian of C_3H_3

We refer to the following equation again:

$$E = E_0 + \langle\Psi_0|\left(\frac{\partial U}{\partial Q}\right)_0|\Psi_0\rangle Q + \frac{1}{2}\langle\Psi_0|\left(\frac{\partial^2 U}{\partial Q^2}\right)_0|\Psi_0\rangle Q^2 + \sum_{k=1}^{\infty}\frac{|\langle\Psi_0|\left(\frac{\partial U}{\partial Q}\right)_0|\Psi_k\rangle|^2 Q^2}{E_0 - E_k}. \tag{3.11}$$

The second term can lead to the linear Jahn–Teller effect causing the instability. For example, let us look at C_3H_3 with the high-symmetry geometry D_{3h}. We can find its ground electronic state of being doubly degenerate, E''. In such a case, the second term contributes to the symmetry-lowering distortion in the ground state via the JT effect. More specifically, the double degeneracy in the ground E'' state is subject to the JT problem $E'' \times (a_1' + e')$. Because the ground E'' state gives rise to $\langle\Psi_0\left(\frac{\partial U}{\partial Q}\right)_0|\Psi_0\rangle \neq 0$ for the vibrational modes of A_1' and E', owing to $[E''^2] = A_1' + E'$ (see Table 5.10). The vibrational modes of A_1' and E' are called the vibronically active modes among the 12 vibrational modes in Fig. 3.12. In par-

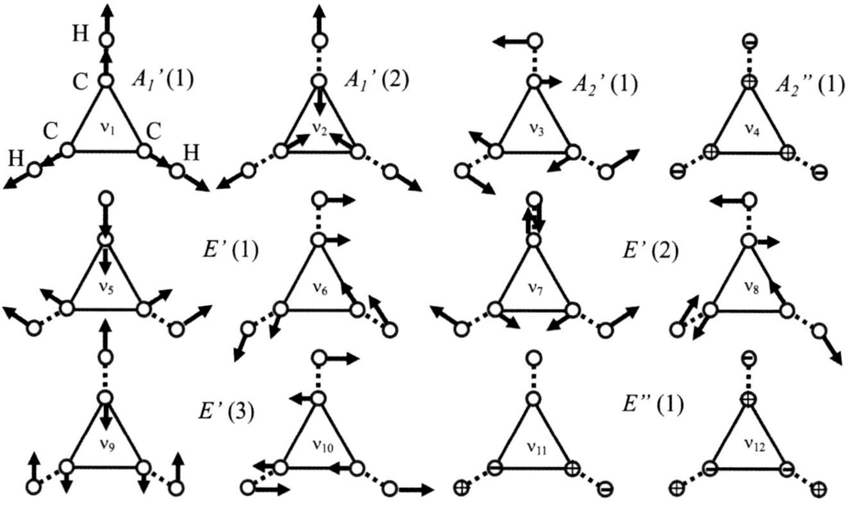

C_3H_3 (D_{3h})

Fig. 3.12 Vibrational normal modes of C_3H_3 (D_{3h})

ticular, the deformations along the E' modes split the doubly degenerate ground E'' state (JT effect), and hence the E' modes are called the JT active modes. The JT effect shortens one of the C–C bonds to be ethylenic, resulting in C_{2v} symmetry.

The VCD analysis gives us more detailed picture than the introductory qualitative discussion. First, we take the non-degenerate $C_3H_3^+$ as the reference system with the high-symmetry geometry of D_{3h}. The geometry optimization and frequency analysis for the reference system were performed at the RHF/STO-3G level of theory. For the target system, i.e., C_3H_3, we employed the state-averaged CASSCF method[3] including σ-orbitals as well as π-orbitals in the active space, and calculated the VCCs from the energy gradients [9].

Table 3.1 summarizes the calculated VCCs for the vibronically active modes, $2A_1' + 3E'$, with the reorganization energies in meV. The reorganization energy E_A corresponds to the relaxation energy relative to the reference geometry along with the A_1' modes, as shown in Fig. 3.13, and E_{JT} corresponds to the stabilization energy due to the JT effect, as shown in Fig. 3.14. The VCCs are represented by V_α's in Table 3.1 in the unit of 10^{-4} a.u. Among the JT active modes, the $E'(2)$ mode has the maximum coupling, whereas the $A_1'(2)$ mode is the strongest in the totally symmetric modes. It was confirmed that the calculated VCCs satisfy the relation of the Clebsch–Gordan coefficients. The negative value of V_α, -9.22, for Ψ_θ and the positive one

[3]CASSCF is the short form of the complete active space self-consistent field. It is one of the post Hartree–Fock methods, i.e., a specific type of electronic state calculation method including correlation effects, in which electronic excited configurations are allowed to be mixed into the ground one by the configuration interaction approach to describe the ground and excited electronic states at the higher level than the Hartree–Fock one.

Table 3.1 Frequencies, ω_α (cm^{-1}), vibronic coupling constants of C_3H_3, V_α (10^{-4} a.u.), the stabilization energy over all the A'_1 modes (E_A), and that over all the E' modes (E_{JT}) (meV) [9]. Reprinted from J. Phys.: Conf. Ser., vol. 428, T. Sato, M. Uejima, N. Iwahara, N. Haruta, K. Shizu, and K. Tanaka, Vibronic coupling density and related concepts, 012010 (2013)

Mode	ω_α /cm^{-1}	V_α /10^{-4}	a.u.
		Ψ_θ	Ψ_ϵ
$A'_1(1)$	1913	−9.83	−9.83
$A'_1(2)$	3758	−13.63	−13.63
$E'(1)\theta$	1060	−2.83	2.83
$E'(2)\theta$	1480	−9.22	9.22
$E'(3)\theta$	3671	−4.60	4.60
E_A /meV		259	259
E_{JT} /meV		311	311

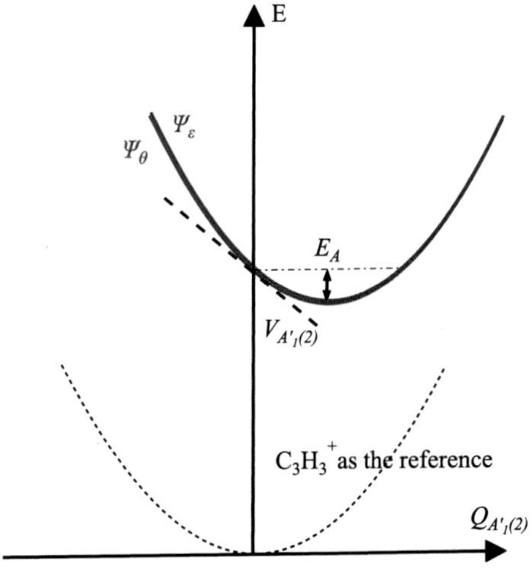

Fig. 3.13 Energy potential surface of C_3H_3 along the coordinate $A'_1(2)$

of V_α, 9.22, for Ψ_ϵ means the negative and positive slopes of the potentials at the reference geometry, as shown in Fig. 3.14. On the other hand, both negative values of V_α, −13.63, for Ψ_θ and Ψ_ϵ represent the negative slopes, as shown in Fig. 3.13.

Figure 3.15 shows the $e''\theta$ LUMO density $\rho_{LU\theta}$ of $C_3H_3^+$ and the electron-density difference $\Delta\rho$ between $C_3H_3^+$ and C_3H_3. Since the LUMO is a π-orbital, the positive LUMO density is distributed outside the molecular plane. On the other hand, $\Delta\rho$ is different from $\rho_{LU\theta}$, and the negative contribution that originates from the polarization of doubly occupied σ-orbitals appears in the molecular plane. The VCD of the strongest mode $A'_1(2)$ is shown in Fig. 3.16 and that of the mode $E'(2)$ with the maximum JT coupling is shown in Fig. 3.17. Since the displacements of all the vibronically active modes are in the molecular plane, and the σ polarization yields large contributions to the VCDs. Figure 3.18 shows the AVCCs of carbon atoms and

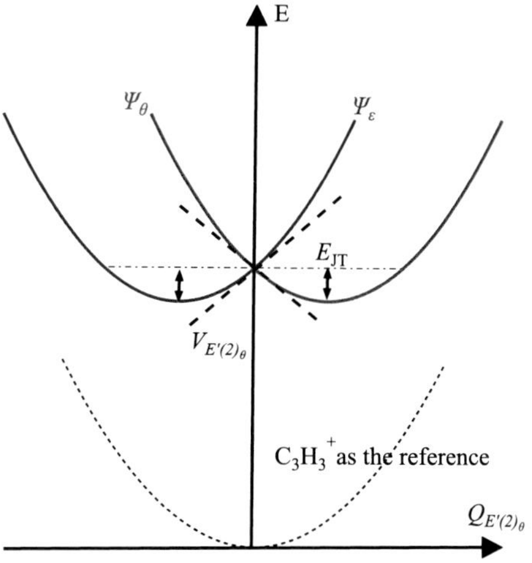

Fig. 3.14 Energy potential surface of C_3H_3 along the coordinate $E'(2)\theta$

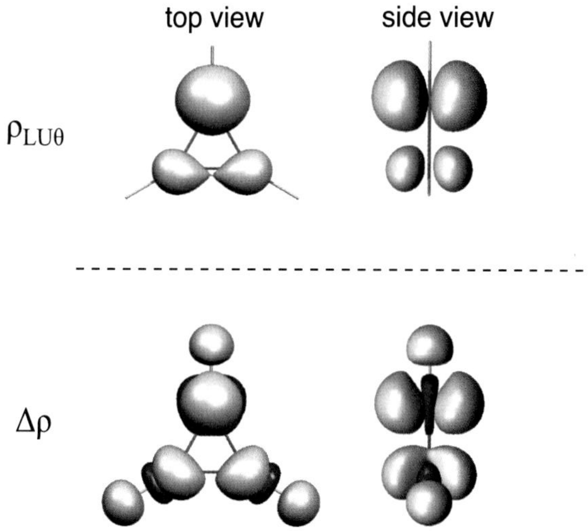

Fig. 3.15 The $e''\theta$ LUMO density $\rho_{LU\theta}$ of $C_3H_3^+$ and $\Delta\rho$ between $C_3H_3^+$ and C_3H_3. The gray region is positive, and the blue region is negative. The isosurface value is 0.01 a.u. [9]. Reprinted from J. Phys.: Conf. Ser., vol. 428, T. Sato, M. Uejima, N. Iwahara, N. Haruta, K. Shizu, and K. Tanaka, Vibronic coupling density and related concepts, 012010 (2013)

the VCD for the $E'(2)$ mode. The negative σ VCD gives rise to the large AVCC on one of the carbon atoms. The VCC calculation with σ polarization allowed is indispensable for evaluating the JT deformation from the equilateral triangle geometry.

From the obtained VCCs, we can estimate the mass-weighted displacement Q_α of the structural relaxation along mode α:

$$Q_\alpha = \frac{|V_\alpha|}{\omega_\alpha^2}. \tag{3.12}$$

It is of importance to remark that Q_α depends on ω_α as well as V_α. It can be intuitively understood as follows: a smaller ω_α means a larger amplitude motion, yielding a

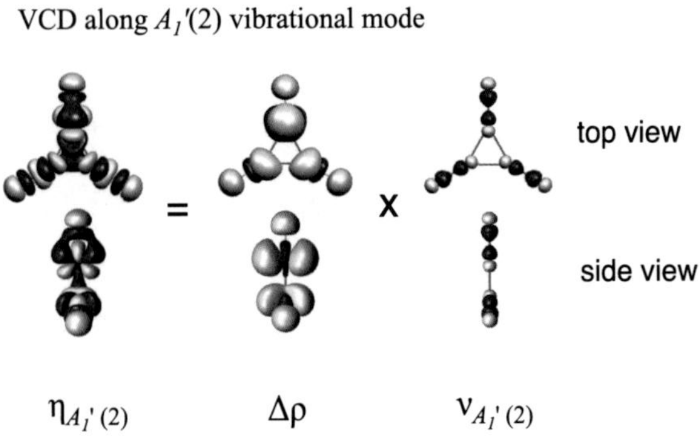

Fig. 3.16 Vibronic coupling density $\eta_{A_1'(2)}$ along $A_1'(2)$ mode = electron-density difference $\Delta\rho$ $\times$ potential derivative $v_{A_1'(2)}$. The isosurface value of $\eta_{A_1'(2)}$ is 0.00005 a.u. [9]. Reprinted from J. Phys.: Conf. Ser., vol. 428, T. Sato, M. Uejima, N. Iwahara, N. Haruta, K. Shizu, and K. Tanaka, Vibronic coupling density and related concepts, 012010 (2013)

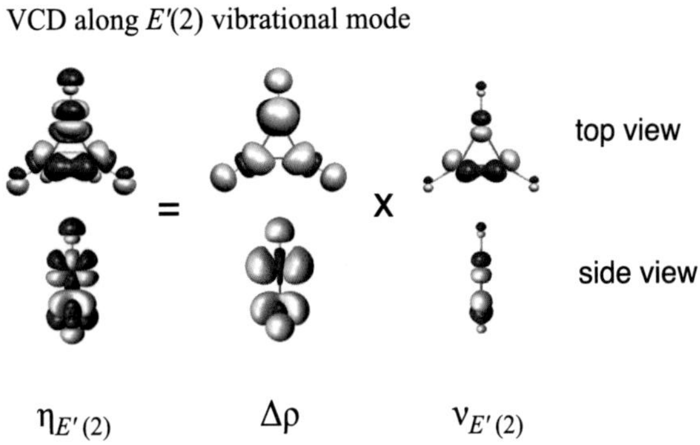

Fig. 3.17 Vibronic coupling density $\eta_{E'(2)}$ along $E'(2)$ mode = electron-density difference $\Delta\rho$ $\times$ potential derivative $v_{E'(2)}$. The isosurface value of $\eta_{E'(2)}$ is 0.00005 a.u. [9]. Reprinted from J. Phys.: Conf. Ser., vol. 428, T. Sato, M. Uejima, N. Iwahara, N. Haruta, K. Shizu, and K. Tanaka, Vibronic coupling density and related concepts, 012010 (2013)

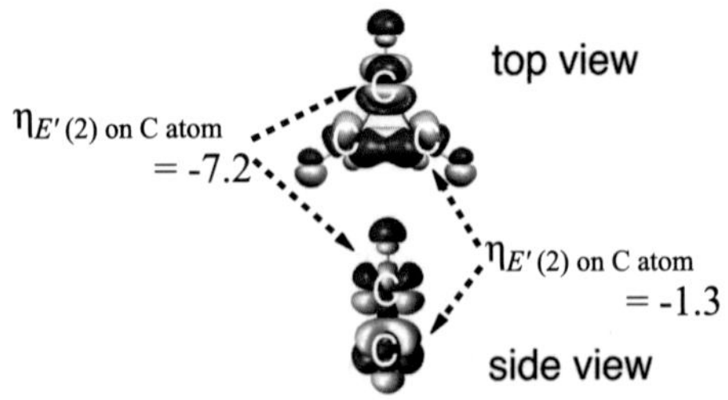

Fig. 3.18 Atomic vibronic coupling constants (10^{-4} a.u.) and vibronic coupling density for the $E'(2)$ mode [9]. Reprinted from J. Phys.: Conf. Ser., vol. 428, T. Sato, M. Uejima, N. Iwahara, N. Haruta, K. Shizu, and K. Tanaka, Vibronic coupling density and related concepts, 012010 (2013)

Table 3.2 Displacements of the structural relaxation of C_3H_3 along all the active modes. Here, ω_α means a frequency, μ_α denotes a reduced mass, V_α stands for a vibronic coupling constant, and q_α is a displacement for mode α

Mode	ω_α /cm^{-1}	μ_α /amu	V_α /10^{-4} a.u.	q_α /Å
$A_1'(1)$	1913	4.37	−9.83	0.077
$A_1'(2)$	3758	1.18	−13.63	0.053
$E'(1)\theta$	1060	1.09	−2.83	0.144
$E'(2)\theta$	1480	5.61	−9.22	0.106
$E'(3)\theta$	3671	1.11	−4.60	0.019

greater structural relaxation. Further, Q_α can be converted to the real displacement q_α:

$$q_\alpha = \frac{Q_\alpha}{\sqrt{\mu_\alpha}}, \tag{3.13}$$

where μ_α is the reduced mass of mode α. For all the active modes of C_3H_3, q_α's are listed in Table 3.2. The asymmetric deformational modes of the carbon triangle framework, $E'(1)$ and $E'(2)$ exhibit large displacements in the order of 10^{-1} Å.

3.6 Vibronic Coupling Density Picture of Diels–Alder Reaction

Molecular deformations play a crucial role in the course of a chemical reaction. In this section, we take Diels–Alder reactions as an example.

Diels–Alder reactions can be well interpreted based on the frontier orbital theory [10, 11], as mentioned for the reaction system of ethylene and butadiene in Sect. 1.6. The VCD picture of the Diels–Alder reaction of butadiene and styrene is introduced here (see Fig. 3.19).

Here, we define the effective vibronic coupling density (EVCD), as defined in Sect. 5.4.4.4, to clarify the regioselectivity of the Diels–Alder reaction. $\eta_s(\boldsymbol{x})$ is the VCD of reaction mode s along which the chemical reaction occurs, and reaction mode is taken as an effective mode that

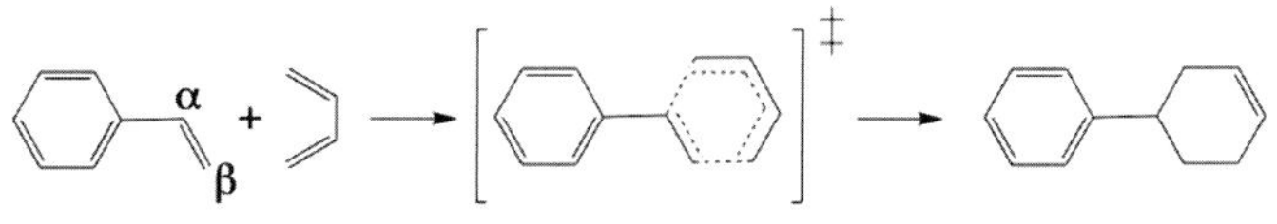

Fig. 3.19 Diels–Alder reaction of butadiene and styrene

$$ds = \sum_{\alpha} \frac{V_{\alpha}}{\sqrt{\sum_{\alpha} V_{\alpha}^2}} dQ_{\alpha}, \tag{3.14}$$

coincides with the steepest descent direction to the minimum of the potential energy surface in the charge-transfer state. Here the V_{α} is VCC along the normal mode α, then ds can be regarded as the normal coordinate deformation weighted with VCC. η_s is represented as the product of electron-density difference due to charge transfer and potential derivative in terms of s. The corresponding VCC is obtained from the gradient of the potential energy surface in the charge-transfer state concerning s. It is noted that η_s includes the effects of not only electronic but also vibrational structures on chemical reactivity as discussed in Sect. 5.4.5.

The results calculated at B3LYP/STO-3G level of theory show that the reactive site of the reaction is the $C_{\alpha} = C_{\beta}$ double bond. The LUMO of styrene is delocalized as shown in Fig. 3.20. The VCD analysis for the effective mode of styrene anion is obtained as shown in Fig. 3.20. The EVCD distribution on C_{α} and C_{β} is fit to that of ethylene as shown in Fig. 3.20, which causes the reacting system to further stabilized through a structural relaxation due to vibronic couplings.

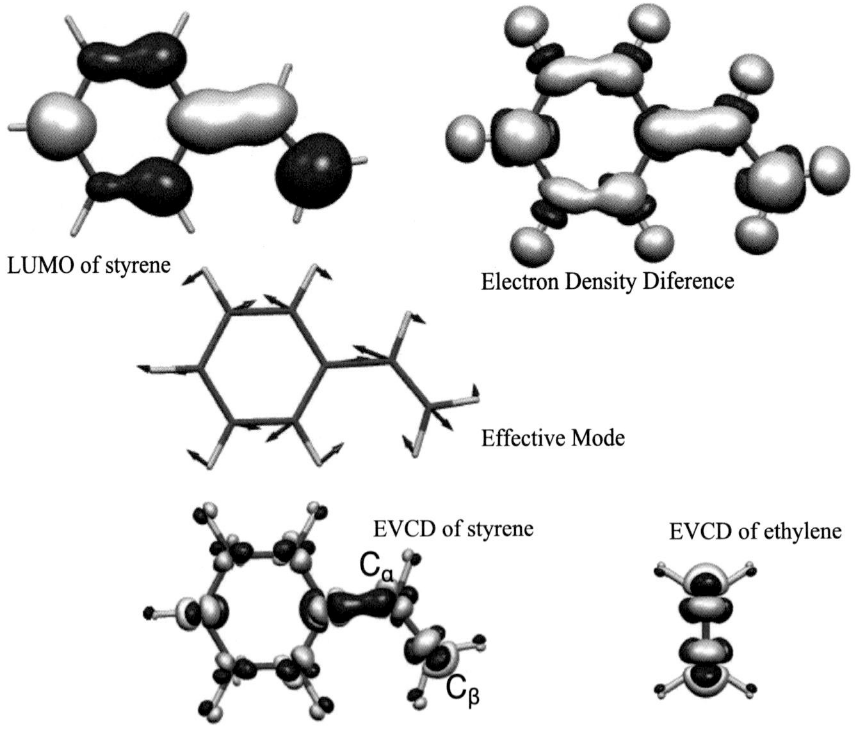

Fig. 3.20 LUMO, electron-density difference, effective mode, and EVCD of styrene and ethylene

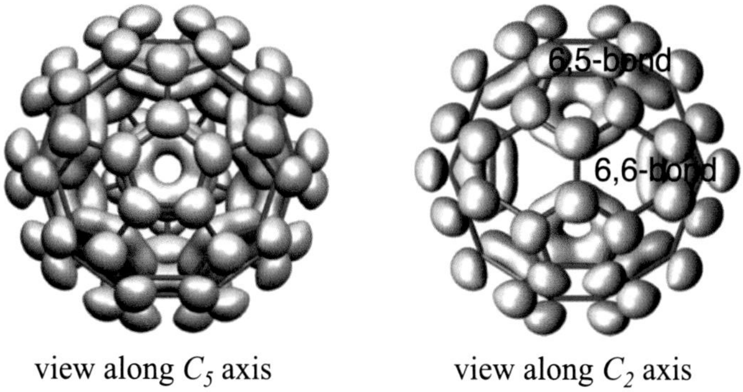

Fig. 3.21 Averaged density of LUMOs

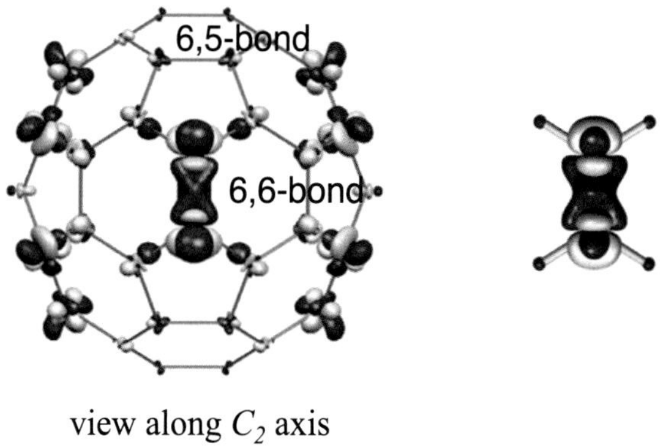

Fig. 3.22 Vibronic coupling density analysis for the reaction mode in C_{60}^- and ethylene monoanion

The VCD as a reactivity index is applied to investigate the regioselectivity of the cycloadditions of fullerene [12–15], metallofullerene [16], and polycyclic aromatic hydrocarbons [17]. C_{60} fullerene undergoes nucleophilic cycloadditions at the 6,6-bonds between two hexagonal rings rather than the 6,5-bonds between the hexagonal and pentagonal rings [18, 19]. The frontier orbital theory has difficulty in predicting this regioselectivity of the reactions because the frontier orbitals tend to be delocalized in such large systems, as shown in Fig. 3.21. Since the LUMO density is delocalized in the entire molecule, the reactive sites in C_{60} cannot be identified by the frontier orbital theory. In contrast, Fig. 3.22 shows the VCD of C_{60}^- concerning the effective mode, which is strongly localized on the 6,6-bonds. Therefore, the VCD elucidates the reactive sites of C_{60}. In addition, the VCD of C_{60}^- is analogous to that of ethylene mono-anion, which implies that C_{60} undergoes nucleophilic reactions in a similar manner to ethylene. The reactive sites for multiple cycloadditions to C_{60} are obtained in agreement with experiments by successively calculating the VCD of C_{60} adducts [14, 15].

The importance of vibronic coupling in chemical reactions can also be corroborated by comparing reaction profiles without and with structural relaxation

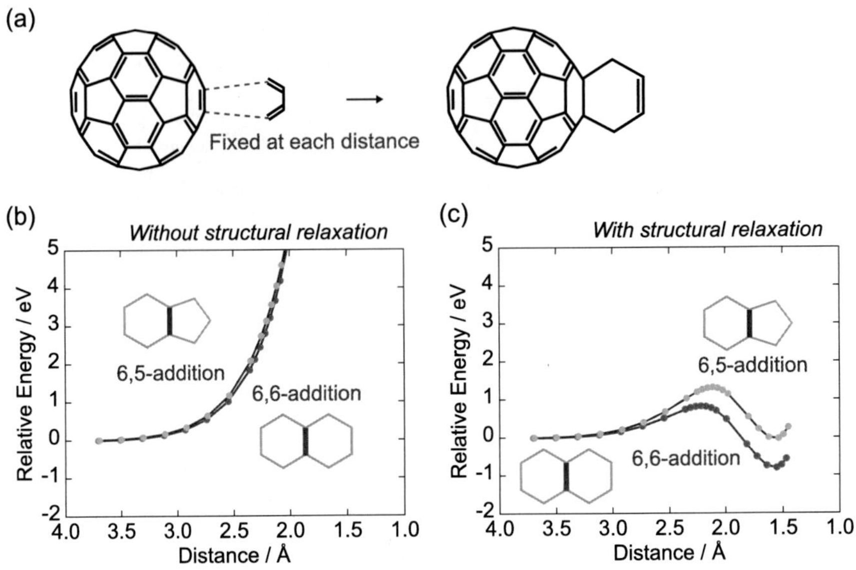

Fig. 3.23 The Diels–Alder reaction of C_{60} + butadiene: **a** its scheme and its energy profiles **b** without and **c** with structural relaxation at the B3LYP/6-311G(d,p) level of theory. Structural relaxation was taken into account by optimizing the total system consisting of the isolated structures of the reactants at each distance between the reaction centers

(Fig. 3.23). As shown in Fig. 3.23b, if structural relaxation is ignored in the Diels–Alder reaction of C_{60} + butadiene, 6,6- and 6,5-additions have no reaction barriers and make no difference in their potential curves. It means that its regioselectivity cannot be described only by charge-transfer interaction (orbital interaction), i.e., the frontier orbital theory. As shown in Fig. 3.23c, however, the consideration of structural relaxation makes reaction barriers in both the pathways and clearly illustrates the lower barrier of the 6,6-addition, which is consistent with the experimental fact. In such a case, vibronic coupling, which is the origin of structural relaxation, is indispensable for describing the regioselectivity. It is the reason for the effectiveness of the VCD analysis in the prediction of chemical reactions.

References

1. Sato T, Tokunaga K, Tanaka K (2006) J Chem Phys 124
2. Sato T, Tokunaga K, Tanaka K (2008) J Phys Chem A 112:758
3. Sato T, Tokunaga K, Iwahara N, Shizu K, Tanaka K (2009) The Jahn–teller effect: fundamentals and implications for physics and chemistry. In: Köppel H, Yarkony DR, Barentzen H (eds). Springer series in chemical physics, vol 97. Springer, Berlin
4. Sato T, Uejima M, Iwahara N, Haruta N, Shizu K, Tanaka K (2013) J Phys Conf Ser 428
5. Renner R (1934) Z Physik 92:172
6. Bersuker I, Gorinchoi N, Polinger V (1984) Theo Chim Acta 66:161

7. Kameoka Y, Sato T, Koyama T, Tanaka K, Kato T (2014) Chem Phys Lett 598:69
8. Kayahara E, Kouyama T, Kato T, Takaya H, Yasuda N, Yamago S (2013) Angew Chem Int Ed 52:13722
9. Sato T, Uejima M, Iwahara N, Haruta N, Shizu K, Tanaka K (2013) J Phys Conf Ser 428(1)
10. Fukui K (1971) Acc Chem Res 4:57
11. Ota W, Sato T, Tanaka K (2018) J Phys Conf Ser 1148
12. Sato T, Iwahara N, Haruta N, Tanaka K (2012) Chem Phys Lett 531:257
13. Haruta N, Sato T, Tanaka K (2012) J Org Chem 77:9702
14. Haruta N, Sato T, Iwahara N, Tanaka K (2013) J Phys: Conf Ser 428(1)
15. Haruta N, Sato T, Tanaka K (2014) Tetrahedron 70:3510
16. Haruta N, Sato T, Tanaka K (2015) J Org Chem 80:141
17. Haruta N, Sato T, Tanaka K (2015) Tetrahedron Lett 56:590
18. Yurovskaya M, Trushkov I (2002) Russ Chem Bull Int Ed 51:367
19. Thilgen C, Diederich F (2006) Chem Rev 106:5049

Chapter 4
Design for Functional Molecules by Vibronic Coupling Density

Abstract The vibronic coupling constant (VCC) offers quantitative insight into the molecular design. The diagonal VCC is the gradient of a potential energy surface concerning the normal coordinate at a fixed nuclear configuration as defined in Sect. 5.4.1, which can give the quantitative evaluation of the force which occurred under an electronic transition. The elongation of bond on anionization of hydrogen and ethylene is interpreted by the value of vibronic coupling density (VCD), as shown in Sect. 3.1. The amplitudes of Jahn–Teller deformation for C_3H_3 are also evaluated by the VCCs, which are analyzed based on the VCD concept. As well as the molecular deformations, the VCD analyses explain carrier-transport efficiency and the rate of internal conversions for some molecular systems. We give the further outlooks for novel functional molecules. Through the VCD analyses, highly efficient molecules for carrier transport and light emission are designed.

Keywords Carrier transporting · Light emission · Franck–Condon state · Internal conversion · Transition dipole moment density · Organic light-emitting diode

4.1 Design for Carrier-Transporting Molecules

When a voltage is applied to an electrode–molecule–electrode system shown in the upper picture of Fig. 4.1, carriers can be transported across the molecule interacting with nuclear vibrations. As shown in the lower picture of Fig. 4.1, the molecule is once ionized by the excitation with $\Delta E_{\mathrm{voltage}}$ and neutralized with an electron through the electrode. The carrier-transporting process across the molecule can be regarded as the ionizations, or electronic excitation with the applied voltage, followed by the vibrational relaxation. Since the vibronic coupling causes heat dissipation, developing a molecule with a small vibronic coupling is important for efficient carrier transport. Although both intramolecular and intermolecular vibronic coupling can affect efficiency, the intramolecular vibronic coupling is only considered here.

Carbazole derivatives are known as having the high efficiency of hole-transporting [1, 2] comparing with the similar structure molecules of biphenyl and fluorene [3]. Their structures are shown in Fig. 4.2, and their VCC and VCD are

T. Kato et al., *Vibronic Coupling Density*,
SpringerBriefs in Molecular Science,
https://doi.org/10.1007/978-981-16-1796-6_4

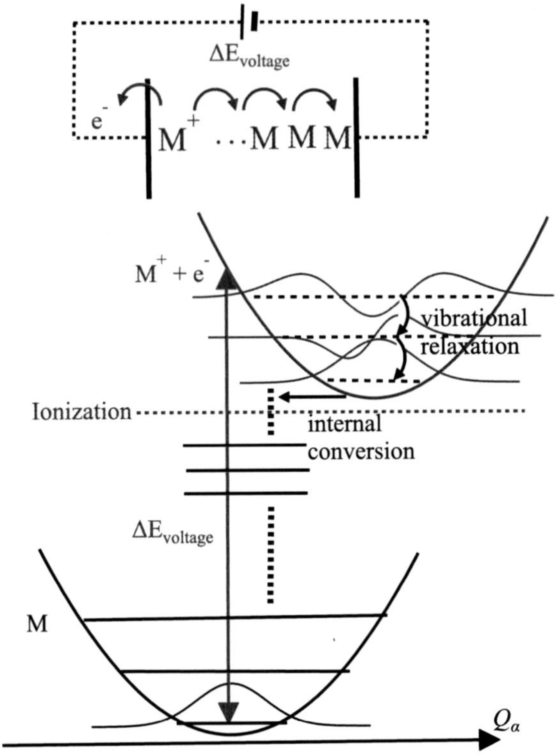

Fig. 4.1 Upper: the hole-transporting process in a electrode–molecule–electrode system with the voltage $\Delta E_{\text{voltage}}$. Lower: the hole-transporting process across a molecule

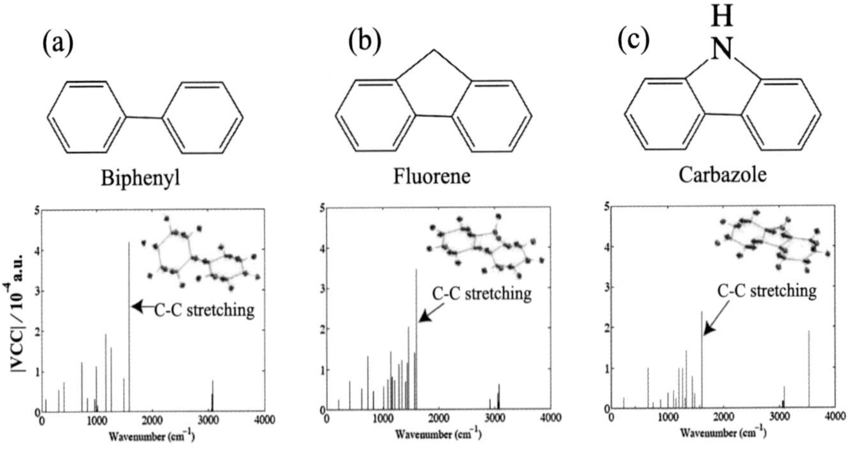

Fig. 4.2 Molecular structures and the absolute values of vibronic coupling constants of **a** biphenyl, **b** fluorene, and **c** carbazole cations. Reprinted from J. Phys.: Conf. Ser., vol. 1148, W. Ohta, T.Sato, and K.Tanaka, Applications of Vibronic Coupling Density, 012004(2018)

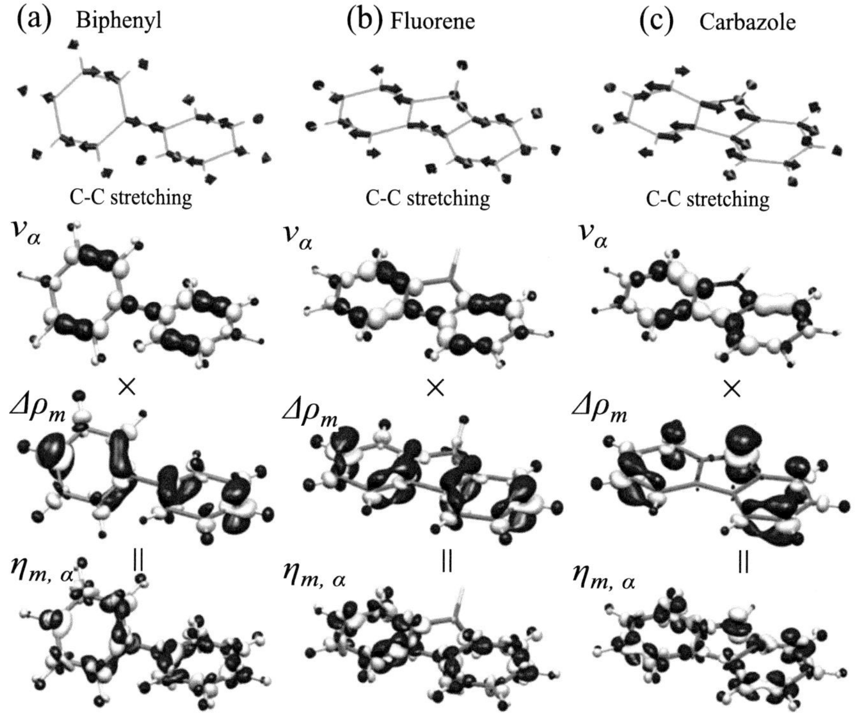

Fig. 4.3 C–C stretching vibrational modes are depicted on the first row for **a** biphenyl, **b** fluorene, and **c** carbazole. The second row shows their derivatives of the electronic–nuclear potential derivatives v_α, the third row the electron-density differences due to ionization $\Delta\rho_m$, and the lowest row are the vibronic coupling densities of the C–C stretching modes $\eta_{m,\alpha}$. The values of $\eta_{m,\alpha}$ are obtained as the product of v_α and $\Delta\rho_m$ along each column. White regions denote positive; blue regions are negative. Reprinted from J. Phys.: Conf. Ser., vol. 1148, W. Ohta, T.Sato, and K.Tanaka, Applications of Vibronic Coupling Density, 012004(2018)

evaluated to clarify why the carbazole is special. Figure 4.2 shows the diagonal VCCs of biphenyl, fluorene, and carbazole cations concerning the strongest coupling modes. In these molecules, the vibrational mode that gives the largest VCC is the C–C stretching mode. The largest VCC values are calculated as -4.192×10^{-4} a.u. for biphenyl, -3.440×10^{-4} a.u. for fluorene, and -2.390×10^{-4} a.u. for carbazole.[1] The VCC of carbazole is smaller than that of biphenyl and fluorene, which indicates that intramolecular vibronic coupling is the weakest in carbazole. Figure 4.3 shows the C–C stretching modes and their potential derivatives v_α. Biphenyl, fluorene, and carbazole exhibit similar distributions of v_α with positive and negative

[1]The a.u. of mass is m_e (electron mass: 0.0005486 amu), a.u. of length a_0 (Bohr radius: 0.5292 Å), and a.u. of energy E_h (Hartree energy: 27.21 eV). VCC and OVCC given here are the mass-weighted constant in $E_h m_e^{-1/2} a_0^{-1}$, which is converted to the value of $2.195 \times 10^3 \times \sqrt{M}$ eV / Å, where M is the reduced mass of each vibrational mode in amu.

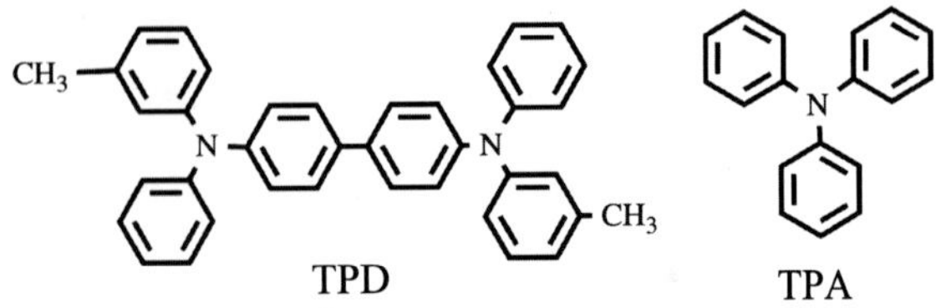

Fig. 4.4 Molecular structures of TPD and TPA

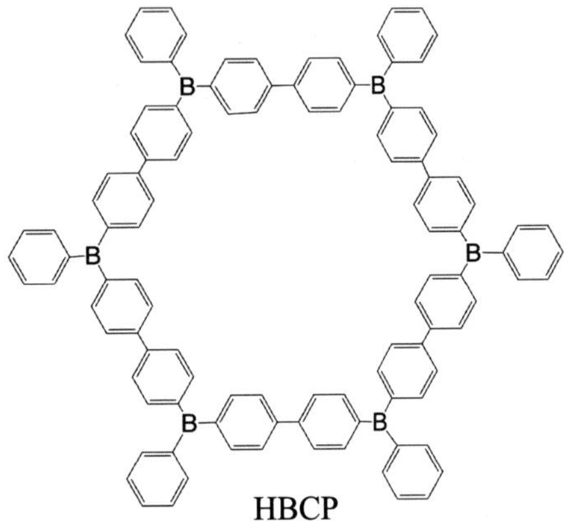

Fig. 4.5 Molecular structure of HBCP

phases symmetrically around the C atoms. Figure 4.3 shows the electron-density differences $\Delta\rho_m$ due to cationization and the diagonal VCD $\eta_{m,\alpha}$. $\Delta\rho_m$ is extensively distributed on benzene rings in biphenyl and fluorene, whereas the distribution on benzene rings is relatively suppressed in carbazole. Also, $\Delta\rho_m$ is localized on the N atom in carbazole, where v_α has small values on the N atom. As shown in Sect. 5.4.1, the diagonal VCD $\eta_{m,\alpha}$ is given as the product of v_α and $\Delta\rho_m$, refer Eq. 5.144. The diagonal VCD $\eta_{m,\alpha}$ decreases for carbazole, resulting in a small VCC. It is the reason why carbazole exhibits high hole-transporting efficiency. The same localization of $\Delta\rho_m$ on a N atom is also responsible for the weak vibronic coupling in N,N'-diphenyl-N,N'-di(m-tolyl)benzidine (TPD) [4] and triphenylamine (TPA) [5], which are shown in Fig. 4.4.

Based on the same consideration using the VCD analysis, hexaboracyclophane (HBCP) [6] and hexaaza[1_6]parabiphenylophane (HAPBP) [7] have been proposed as highly efficient carrier-transporting molecules with a weak vibronic coupling (see Figs. 4.5 and 4.6).

Fig. 4.6 Molecular structure of HAPBP

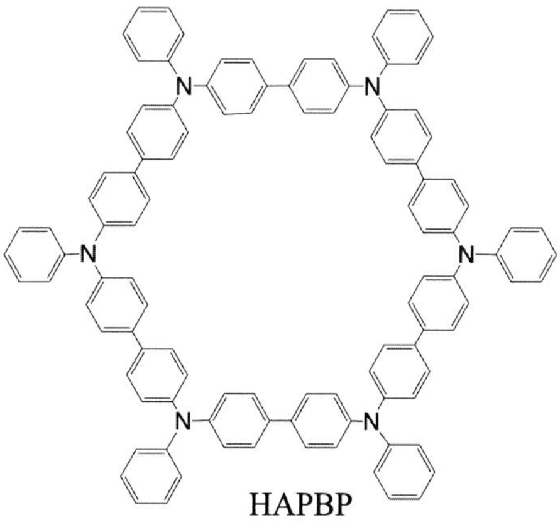

HAPBP

4.2 Design for Light-Emitting Molecules

Theoretical design of highly efficient light-emitting molecules is of interest for application to organic light-emitting diodes. One of the essential factors for efficiency is the fluorescence quantum yield, determined by the radiative and non-radiative processes' ratio. Therefore, the molecule should be designed to enhance the radiative process and the non-radiative process should be suppressed. The non-radiative process includes vibrational relaxation, internal conversion, and intersystem crossing. Here we discuss vibrational relaxation processes and the internal conversion and give their relations with the vibronic coupling. The radiative process is also discussed because the transition dipole moment density (TDMD) can be considered in a similar framework as the VCD described in Sect. 5.4.3. Furthermore, the radiative rate constant depends on a Franck–Condon factor, calculated using diagonal VCCs.

Figure 4.7 describes the photoexcitation process followed by the vibrational relaxation and internal conversion. Assuming that nuclei's motions are slow compared with that of electrons, a vertical transition occurs from a ground S_0 state to a Franck–Condon state by the absorption of a photon. A molecule at the Franck–Condon state transforms into an adiabatic state through vibrational relaxation. The molecule is stabilized due to vibrational relaxation by dissipating the vibrational energy to the surrounding environment. Within the crude adiabatic approximation, the potential energy surface at the S_m state is represented as [8]

$$E_m(\mathbf{R}) = E_m(\mathbf{R}^0) + \sum_{\alpha} \left\{ \frac{\omega_{m,\alpha}^2}{2} \left(Q_\alpha - \frac{|V_{m,\alpha}|}{\omega_{m,\alpha}^2} \right)^2 - \frac{V_{m,\alpha}^2}{2\omega_{m,\alpha}^2} \right\}, \tag{4.1}$$

where $\omega_{m,\alpha}$ is the frequency of vibrational mode α. The minima of the energy surface for each vibrational mode are stabilized by $-V_{m,\alpha}^2/2\omega_{m,\alpha}^2$ with the shift of $V_{m,\alpha}/\omega_{m,\alpha}^2$ in the direction of Q_α. The total stabilization energy due to the vibrational relaxation is obtained by

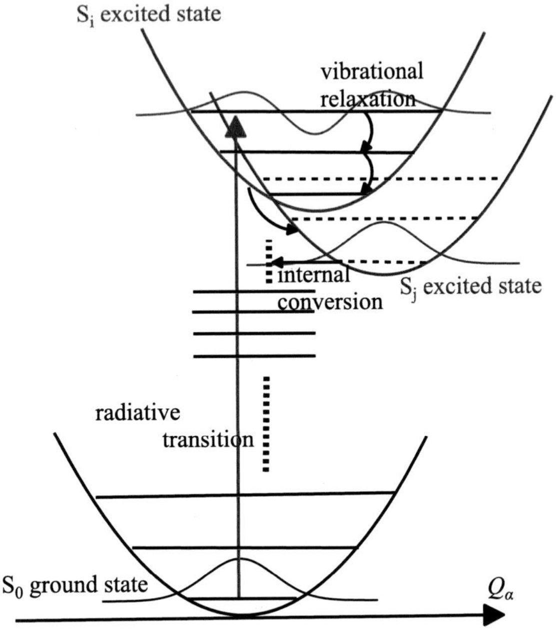

Fig. 4.7 The process of the photoexcitation followed by the vibrational relaxation and the internal conversion

$$\Delta E_m^{\text{stab}} = \sum_\alpha \frac{V_{m,\alpha}^2}{2\omega_{m,\alpha}^2}. \tag{4.2}$$

Thus, vibrational relaxation can be reduced by decreasing the values of diagonal VCC. Here the normal modes in the excited state are assumed to be the same as that in the ground state.[2]

Since vibrational relaxation is a rapid process, radiative transition and internal conversion mostly occur at the adiabatic state. Assuming that the internal conversion occurs from $|\Phi_{mv}\rangle = |\chi_{mv}\rangle|\Psi_m\rangle$ to $|\Phi_{nv'}\rangle = |\chi_{nv'}\rangle|\Psi_n\rangle$, the rate constant of internal conversion from S$_m$ to S$_n$ states is given by [8]

$$k_{m\to n}^{\text{IC}}(T) = \frac{2\pi}{\hbar} \sum_{\alpha,\beta} V_{mn,\alpha} V_{nm,\beta} \sum_{v,v'} P_{mv}(T)\rho(E_{mv}) \times \langle\chi_{mv}|Q_\alpha|\chi_{nv'}\rangle\langle\chi_{nv'}|Q_\beta|\chi_{mv}\rangle, \tag{4.3}$$

where $P_{mv}(T)$ is the Boltzmann distribution at $|\Phi_{mv}\rangle$ with temperature T, and E_{mv} and $E_{nv'}$ are the eigenvalues corresponding to $|\Phi_{mv}\rangle$ and $|\Phi_{nv'}\rangle$, respectively. $\rho(E)$ stands for the density of states in the final state $|\Phi_{nv'}\rangle$. In Eq. (4.3), the term involving

[2]The Duschinsky effect is neglected in Eq. (4.2). The stabilization energy due to the Duschinsky effect is evaluated from the quadratic VCC [8]. Within the Born–Oppenheimer approximation, the stabilization energy including the Duschinsky effect is calculated from the difference in the energies of the Franck–Condon S$_m$ state and S$_m$ state at an equilibrium nuclear configuration.

two vibrational modes is neglected. Thus, the internal conversion can be suppressed by decreasing the values of off-diagonal VCC. $k^{\mathrm{IC}}_{m\to n}$ also depends on the diagonal VCC through $\langle \chi_{nv'}|Q_\alpha|\chi_{mv}\rangle$. The initial vibrational state $|\chi_{mv}\rangle$ is expressed as a product of the independent initial states $|v_\alpha\rangle$ with the number of phonons v_α. Similarly, the final vibrational state $|\chi_{nv'}\rangle$ is expressed as a product of the independent final states $|v'_\alpha\rangle$ with the number of phonons v'_α. As a result,

$$\langle \chi_{nv'}|Q_\alpha|\chi_{mv}\rangle = \langle v'_\alpha|Q_\alpha|v_\alpha\rangle \prod_{\beta\neq\alpha} \left\langle v'_{n,\beta}\middle| v_{m,\beta}\right\rangle \tag{4.4}$$

with [9]

$$\langle v'_\alpha||v_\alpha\rangle = \sqrt{\frac{v_\alpha! v'_\alpha!}{2^{v_\alpha+v'_\alpha}}}\, e^{-\frac{1}{4}g^2_{m,\alpha}} \sum_{l=0}^{\min[v_\alpha, v'_\alpha]} (-1)^{v'_\alpha - l} 2^l \frac{g_{m,\alpha}^{v_\alpha+v'_\alpha-2l}}{l!(v_\alpha - l)!(v'_\alpha - l)!}, \tag{4.5}$$

where $g_{m,\alpha} = V_{m,\alpha}/\sqrt{\hbar\omega^3_{m,\alpha}}$ is the dimensionless VCC. Equation (4.3) is based on the crude adiabatic approximation. The rate constant of internal conversion has been derived in Refs. [10–12] based on the Born–Oppenheimer approximation.

The rate constant of radiative transition from S_m to S_n states is [8]

$$k^{\mathrm{r}}_{m\to n}(T) = \int_0^\infty d\omega \frac{4\omega^3}{3c^3} \sum_{mn} P_{mv}(T)\mu^2_{mn} \left\langle \chi_{mv}\middle|\chi_{nv'}\right\rangle^2 \rho(E_{mv} - \hbar\omega), \tag{4.6}$$

where ω and c are the angular frequency and speed of photon, respectively. μ_{mn} is the transition dipole moment between $|\Psi_m\rangle$ and $|\Psi_n\rangle$, which affects the enhancement of radiative transition rate. The VCD is introduced as the integrand of VCC to control its value based on the local picture. Similarly, the transition dipole moment density (TDMD) is introduced as the transition dipole moment's integrand [13]:

$$\mu_{mn} = \int \tau_{mn}(\boldsymbol{x})\, d^3\boldsymbol{x}. \tag{4.7}$$

τ_{mn} is represented using the overlap density between S_m and S_n states as

$$\tau_{mn}(\boldsymbol{x}) = -e\boldsymbol{x}\rho_{mn}(\boldsymbol{x}), \tag{4.8}$$

where e is the charge of an electron shown in Sect. 5.4.3.

A design principle for fluorescent anthracene derivatives, which are candidates for highly efficient emitting molecules in an organic light-emitting diode (OLED), has been proposed from the analyses of VCD and TDMD [14, 15]. Figure 4.8 shows the molecules of anthracene (**A**), 9-chloroanthracene (**CA**), and 9,10-dichloroanthracene (**DCA**), whose fluorescence quantum yields were experimentally evaluated as 0.33, 0.11, and 0.53, respectively [16]. First, the contribution of vibrational relaxation to the

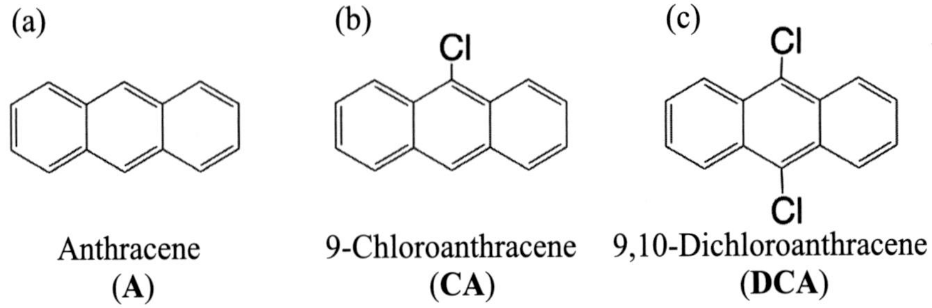

Fig. 4.8 The molecular structures of **a** anthracene (A), **b** 9-chloroanthracene (CA), and **c** 9,10-dichloroanthracene (DCA)

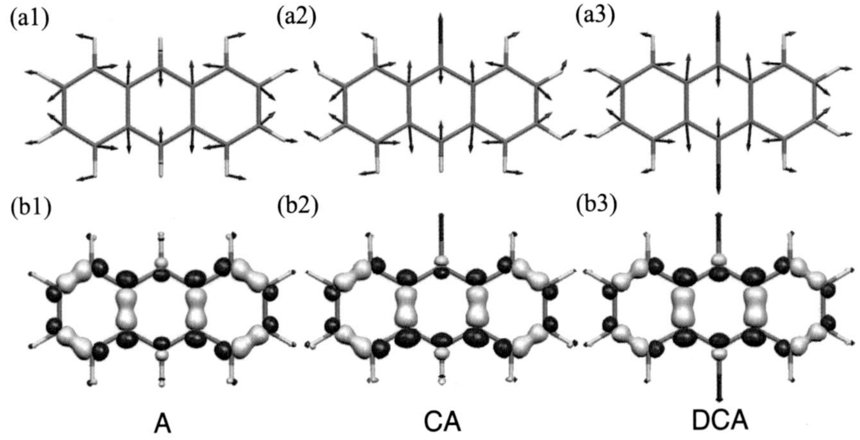

Fig. 4.9 (a1–a3) Vibrational modes susceptible to chlorinations and (b1–b3) derivatives of electronic–nuclear potentials with respect to their modes v_α: (a1, b1) **A** ($\omega_{48} = 1427.55\ \text{cm}^{-1}$), (a2, b2) **CA** ($\omega_{48} = 1420.09\ \text{cm}^{-1}$), and (a3, b3) **DCA** ($\omega_{49} = 1409.13\ \text{cm}^{-1}$). White regions are positive; blue regions are negative. Reprinted with permission from Ref. [14]

quantum yield is discussed [14]. Figure 4.9 shows the vibrational modes of **A**, **CA**, and **DCA** susceptible to chlorinations and the potential derivatives v_α concerning the susceptible modes. The modes and v_α exhibit similar behaviors regardless of chlorinations. Figure 4.10 shows the electron-density differences $\Delta\rho_1$ between S_1 and S_0 states, and the diagonal VCD $\eta_{1,\alpha}$. $\Delta\rho_1$ are large at the C atoms bonded to the Cl atoms in **CA** and **DCA**. As a result, $\eta_{1,\alpha}$, the product of v_α and $\Delta\rho_1$, has large positive values on these C atoms although $\eta_{1,\alpha}$ localized on the edges of **CA** and **DCA** are negative. Therefore, the diagonal VCCs obtained by the integration of diagonal VCD are small in **CA** and **DCA** compared to **A**, which explains the order of the calculated stabilization energies: 0.2332 eV for **A**, 0.2275 eV for **CA**, and 0.2211 eV for **DCA**. Although the vibrational relaxation is suppressed in **CA** and **DCA**, the quantum yield of **CA** is lower than that of **A**. It is attributed to the difference in the internal conversion rate. The square sums of off-diagonal VCC,

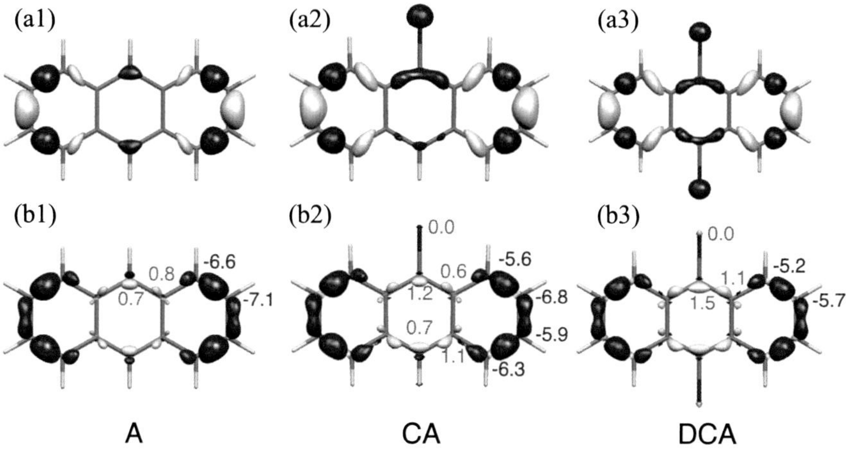

Fig. 4.10 (a1–a3) Electron-density differences between S_1 and S_0 states $\Delta\rho_1$ and (b1–b3) diagonal vibronic coupling densities in the Franck–Condon S_1 state for normal modes susceptible to chlorinations $\eta_{1,\alpha}$: (a1, b1) **A**, (a2, b2) **CA**, and (a3,b3) **DCA**. Atomic vibronic coupling constants in 10^{-5} a.u. are shown in (b1–b3). Large AVCC values are shown in red and those reduced significantly by chlorinations are shown in blue. White regions are positive; blue regions are negative. Reprinted with permission from Ref.[14]

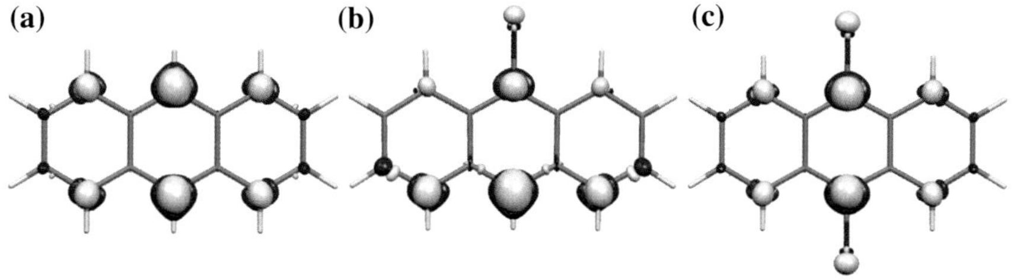

Fig. 4.11 Transition dipole moment densities between S_1 and S_0 states in z-direction $\tau_{10,z}$, for **a A**, **b CA**, and **c DCA**. White regions are positive; blue regions are negative. Reprinted with permission from Ref. [14]

$\sum_\alpha |V_{10,\alpha}|$, are calculated as 1.30×10^{-4} a.u. for **A**, 1.69×10^{-4} a.u. for **CA**, and 1.24×10^{-4} a.u. for **DCA**. According to the selection rule, active vibrational modes that give non-zero VCC are few in a molecule with high symmetry. Therefore, **A** and **DCA** with D_{2h} symmetry have a smaller number of active modes than **CA** with C_{2v} symmetry does, which results in the suppression of internal conversion in **A** and **DCA**. Finally, the TDMD is shown in Fig. 4.11 to evaluate the radiative transition rate. The TDMD is large on Cl atoms because the overlap density ρ_{10} is localized on the Cl atoms. Therefore, the transition dipole moment is increased in **CA** and, even more, in **DCA**. Indeed, the oscillator strengths, proportional to the square of the transition dipole moment, are large in the order of 0.0542 for **A**, 0.0700 for **CA**, and 0.0899 for **DCA**. Since the TDMD is the product of ρ_{10}

Fig. 4.12 The molecular structures of 5,11-bis(phenylethynyl)benzo[1,2-f:4,5-f′]diisoindole-1,3,7,9(2*H*,8*H*)-tetraone

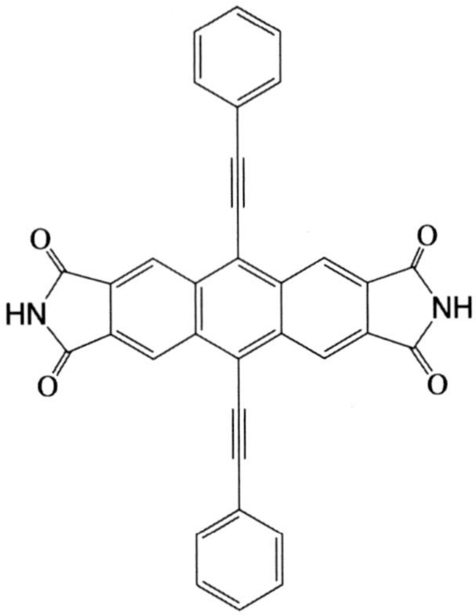

5,11-bis(phenylethynyl)benzo[1,2-f:4,5-f′]diisoindole-1,3,7,9(2*H*,8*H*)-tetraone

and distance from the origin, the localization of ρ_{10} away from the origin leads to the enhancement of radiative transition. Consequently, the design principle is summarized as follows: long substituents on which the overlap density is localized should be introduced, and the D_{2h} symmetry should be maintained. Following these design principles, an anthracene derivative, 5,11-bis(phenylethynyl)benzo[1,2-f:4,5-f′]diisoindole-1,3,7,9(2*H*,8*H*)-tetraone, was designed and synthesized [15], whose structure is shown in Fig. 4.12. The observed fluorescence quantum yield of the rationally designed molecule was 96%.

In addition to the anthracene derivatives, design principles for triphenylamine derivatives were presented [17]. Furthermore, a novel electroluminescence mechanism in OLEDs utilizing higher triplet states than the T_1 state, called fluorescence via higher triplets (FvHT) mechanism, has been proposed based on the analyses of VCD and TDMD [18–20].

References

1. Koene BE, Loy DE, Thompson ME (1998) Chem Mater 10(8):2235
2. Zhang Q, Chen J, Cheng Y, Wang L, Ma D, Jing X, Wang F (2004) J Mater Chem 14(5):895
3. Shizu K, Sato T, Tanaka K, Kaji H (2010) Org Electro 11(7):1277
4. Sato T, Shizu K, Kuga T, Tanaka K, Kaji H (2008) Chem Phys Lett 458(1–3):152
5. Shizu K, Sato T, Tanaka K, Kaji H (2010) Chem Phys Lett 486(4–6):130
6. Shizu K, Sato T, Tanaka K, Kaji H (2010) Appl Phys Lett 97(14)
7. Shizu K, Sato T, Ito A, Tanaka K, Kaji H (2011) J Mater Chem 21(17):6375
8. Uejima M, Sato T, Yokoyama D, Tanaka K, Park JW (2014) Phys Chem Chem Phys 16(27):14244

9. Hutchisson E (1930) Phys Rev 36(3):410
10. Peng Q, Yi Y, Shuai Z, Shao J (2007) J Chem Phys 126(11)
11. Peng Q, Yi Y, Shuai Z, Shao J (2007) J Am Chem Soc 129(30):9333
12. Niu Y, Peng Q, Deng C, Gao X, Shuai Z (2010) J Phys Chem A 114(30):7817
13. Sato T, Uejima M, Iwahara N, Haruta N, Shizu K, Tanaka K (2013) J Phys Conf Ser 428(1)
14. Uejima M, Sato T, Tanaka K, Kaji H (2014) Chem Phys 430:47
15. Uejima M, Sato T, Detani M, Wakamiya A, Suzuki F, Suzuki H, Fukushima T, Tanaka K, Murata Y, Adachi C et al (2014) Chem Phys Lett 602:80
16. Ateş S, Yildiz A (1983) J Chem Soc Faraday Trans 179(12):2853
17. Kameoka Y, Uebe M, Ito A, Sato T, Tanaka K (2014) Chem Phys Lett 615:44
18. Sato T, Uejima M, Tanaka K, Kaji H, Adachi C (2015) J Mater Chem C 3(4):870
19. Sato T (2015) J Comput Chem Jpn 14(6):189
20. Sato T, Hayashi R, Haruta N, Pu YJ (2017) Sci Rep 7(1):4820

Chapter 5
Definitions and Derivations

Abstract The vibronic coupling constant (VCC), as well as the vibronic coupling density (VCD), will be exactly defined in this chapter. Before the precise definition of VCC and VCD, we will introduce Dirac's notation and the basic concept of group and symmetry. The definition of the normal coordinates will be given, and the characters and the direct products will be tabulated for the treatments of group and symmetry.

Keywords Vibronic coupling constant · Vibronic coupling density · Jahn-Teller effect · Hellmann-Feynman theorem · Transition density · Transition dipole moment density · Atomic vibronic coupling constant · Regional vibronic coupling density · Orbital vibronic coupling constant · Orbital vibronic coupling density · Reduced vibronic coupling density · Effective vibronic coupling constant · Effective vibronic coupling density · Fukui function · Nuclear Fukui function · Electron-phonon coupling · Electron-lattice coupling · Rys-Huang factor

5.1 Dirac's Notations

Here we briefly summarize Dirac's notation , bra and ket.

The axioms of quantum mechanics are as follows:

1. (**State**) A quantum-mechanical state $|\psi\rangle$ is a normalized vector in a Hilbert space on the complex field $\mathbb{C}$.
2. (**Observable**) An observable, or physical quantity, $\mathcal{A}$ is expressed by an Hermitian operator $\hat{A}$.
3. (**Measurement**) Any observed value of $\mathcal{A}$ is one of the eigenvalues $\{a_n\}$ of $\hat{A}$.

$$\hat{A}|a_n\rangle = a_n|a_n\rangle. \tag{5.1}$$

4. (**Probability interpretation**) A state $|\psi\rangle$ is expanded by the eigenstates $\{|a_n\rangle\}$ of $\mathcal{A}$:

$$|\psi\rangle = \sum_n |a_n\rangle c_n, \quad c_n \in \mathbb{C}. \tag{5.2}$$

T. Kato et al., *Vibronic Coupling Density*,
SpringerBriefs in Molecular Science,
https://doi.org/10.1007/978-981-16-1796-6_5

When $\mathcal{A}$ is observed, the probability $P(a_n)$ obtaining a_n is

$$P(a_n) = |c_n|^2. \tag{5.3}$$

5. (**Schrödinger equation**) The time dependence of a state $|\psi\rangle$ is described by

$$i\hbar\frac{\partial}{\partial t}|\psi\rangle = \hat{H}|\psi\rangle, \tag{5.4}$$

where $\hat{H}$ is Hamiltonian which is the operator corresponding to energy.

The vector $|\psi\rangle$ is called a *ket*.

The inner product between any kets $|\psi\rangle$ and $|\phi\rangle$ in the Hilbert space is written as $\langle\psi|\phi\rangle$. For complex numbers, $\alpha, \beta \in \mathbb{C}$, therefore,

1. $\langle\phi|\psi\rangle = \langle\psi|\phi\rangle^*$
2. $\langle\varphi|\{\alpha|\phi\rangle + \beta|\phi\rangle\} = \alpha\langle\varphi|\psi\rangle + \beta\langle\varphi|\phi\rangle$
3. $\langle\phi|\phi\rangle \geq 0$ (if $|\phi\rangle = 0$ (zero vector), then $\langle\phi|\phi\rangle = 0$)

where $*$ denotes complex conjugate.

Considering an arbitrary $|\phi\rangle$ for a fixed $|\psi\rangle$, the inner product $\langle\psi|\phi\rangle$ can be regarded as a *function*, $\langle\psi|(|\phi\rangle)$ acting on a ket $|\phi\rangle$ giving a complex number $\langle\psi|\phi\rangle$. The *function* $\langle\psi|$ is called a *bra*.

For an observable $\mathcal{A}$, we have a set of the eigenvectors $\{|a_n\rangle\}$ with the eigenvalues $\{a_n\}$ of the corresponding Hermitian operator $\hat{A}$:

$$\hat{A}|a_n\rangle = a_n|a_n\rangle \tag{5.5}$$

Since $\hat{A}$ is Hermitian, we can take $\{|a_n\rangle\}$ as an orthonormal set:

$$\langle a_m|a_n\rangle = \delta_{mn}. \tag{5.6}$$

n and m can be a continuous variable, x and y. For such a case,

$$\langle a_x|a_y\rangle = \delta(x - y), \tag{5.7}$$

where $\delta(x)$ is Dirac's delta function. Furthermore, we assume that any ket $|\Psi\rangle$ can be expanded in terms of $\{|a_n\rangle\}$:

$$|\Psi\rangle = \sum_n |a_n\rangle c_n, \tag{5.8}$$

where c_n is a complex number. In other words, the basis set $\{|a_n\rangle\}$ is assumed to be complete. Applying $\langle a_m|$ from the left side,

$$\langle a_m|\Psi\rangle = \sum_n \langle a_m|a_n\rangle c_n = \sum_n \delta_{mn} c_n = c_m. \tag{5.9}$$

Substituting (5.9) into Eq. (5.8), the ket $|\Psi\rangle$ can be rewritten as

$$|\Psi\rangle = \sum_n |a_n\rangle\langle a_n|\Psi\rangle. \tag{5.10}$$

Since $|\Psi\rangle$ is arbitrary, we identify

$$\sum_n |a_n\rangle\langle a_n| = \hat{1}, \tag{5.11}$$

where $\hat{1}$ is the unit operator. Equation (5.11) is called the completeness relation. For a continuous case, (5.11) becomes

$$\int dx |a_x\rangle\langle a_x| = \hat{1}. \tag{5.12}$$

For $\hat{B}$, using Eq. (5.11),

$$\hat{B} = \hat{1}\hat{B}\hat{1} = \left(\sum_m |a_m\rangle\langle a_m|\right)\hat{B}\left(\sum_n |a_n\rangle\langle a_n|\right) = \sum_{m,n} |a_m\rangle\langle a_m|\hat{B}|a_n\rangle\langle a_n| = \sum_{m,n} |a_m\rangle B_{mn}\langle a_n|, \tag{5.13}$$

where

$$B_{mn} := \langle a_m|\hat{B}|a_n\rangle \tag{5.14}$$

is called a matrix representation of $\hat{B}$ in terms of the eigenvectors of $\hat{A}$.

A commutator of $\hat{A}$ and $\hat{B}$ is defined by

$$[\hat{A}, \hat{B}] := \hat{A}\hat{B} - \hat{B}\hat{A}. \tag{5.15}$$

For operators $\hat{A}$ and $\hat{B}$ which are commutable; $[\hat{A}, \hat{B}] = 0$, there exist simultaneous eigenvectors such that

$$\hat{A}|a_n, b_n\rangle = a_n|a_n, b_n\rangle, \quad \hat{B}|a_n, b_n\rangle = b_n|a_n, b_n\rangle. \tag{5.16}$$

For example, for an electron in a hydrogen atom, Hamiltonian $\hat{H}$, the square of the orbital angular momentum $\hat{\boldsymbol{L}}^2$, the z-component of the orbital angular momentum $\hat{L}_z$, the square of the spin angular momentum $\hat{\boldsymbol{S}}^2$, and the z-component of the spin angular momentum $\hat{S}_z$ are commutable. Therefore, there exist simultaneous eigenvectors

$$|n, l, m_l, s, m_s\rangle, \tag{5.17}$$

where n is the principle quantum number.

For position operator $\hat{\boldsymbol{r}}$, the eigenequation is written as

$$\hat{\boldsymbol{r}}|\boldsymbol{r}'\rangle = \boldsymbol{r}'|\boldsymbol{r}'\rangle. \tag{5.18}$$

The completeness relation is

$$\int dv_{\boldsymbol{r}}|\boldsymbol{r}\rangle\langle\boldsymbol{r}| = \hat{1}, \tag{5.19}$$

where $dv_{\boldsymbol{r}}$ is the volume element in $\boldsymbol{r}$-space. For an arbitrary ket $|\Psi\rangle$,

$$|\Psi\rangle = \hat{1}|\Psi\rangle = \int dv_{\boldsymbol{r}}|\boldsymbol{r}\rangle\langle\boldsymbol{r}|\Psi\rangle = \int dv_{\boldsymbol{r}}|\boldsymbol{r}\rangle\Psi(\boldsymbol{r}), \tag{5.20}$$

where the coefficient $\Psi(\boldsymbol{r})$ is called the wave function.

Operators are quantized by introducing the canonical commutator relation:

$$[\hat{r}_i, \hat{p}_j] = i\hbar\delta_{ij}, \quad [\hat{r}_i, \hat{r}_j] = 0, \quad [\hat{p}_i, \hat{p}_j] = 0, \tag{5.21}$$

where $\hat{r}_i$ $(i = 1, 2, 3)$ are Cartesian coordinates; $\hat{x}$, $\hat{y}$, $\hat{z}$, describing the position of a particle, and $\hat{p}_i$ $(i = 1, 2, 3)$ are the momenta conjugate to $\hat{r}_i$, $\hat{p}_x$, $\hat{p}_y$, $\hat{p}_z$. It should be noted that Eqs. (5.21) are valid only for Cartesian coordinates and their conjugate momenta. The quantization by Eqs. (5.21) is called canonical quantization.

In the $\boldsymbol{r}$-representation, the position operator $\hat{\boldsymbol{r}}$ is

$$\langle\boldsymbol{r}|\hat{\boldsymbol{r}}|\boldsymbol{r}'\rangle = \boldsymbol{r}'\delta(\boldsymbol{r} - \boldsymbol{r}'). \tag{5.22}$$

Since a function of $\hat{\boldsymbol{r}}$, $f(\hat{\boldsymbol{r}})$ is

$$\langle\boldsymbol{r}|f(\hat{\boldsymbol{r}})|\boldsymbol{r}'\rangle = f(\boldsymbol{r}')\delta(\boldsymbol{r} - \boldsymbol{r}'), \tag{5.23}$$

a potential energy operator consisting of the sum of one-electron operators $V(\hat{\boldsymbol{r}})$ is

$$\langle\boldsymbol{r}|V(\hat{\boldsymbol{r}})|\boldsymbol{r}'\rangle = V(\boldsymbol{r}')\delta(\boldsymbol{r} - \boldsymbol{r}'). \tag{5.24}$$

Using Eq. (5.21), the $\boldsymbol{r}$-representation of the momentum operator $\hat{\boldsymbol{p}}$ is expressed by

$$\langle\boldsymbol{r}|\hat{\boldsymbol{p}}|\boldsymbol{r}'\rangle = -i\hbar\nabla_{\boldsymbol{r}'}\delta(\boldsymbol{r} - \boldsymbol{r}'). \tag{5.25}$$

Accordingly, the Hamiltonian operator $\hat{H}$ in the $\boldsymbol{r}$-representation is diagonal:

$$\langle\boldsymbol{r}|\hat{H}|\boldsymbol{r}'\rangle = \hat{H}(\boldsymbol{r}, \boldsymbol{r}')\delta(\boldsymbol{r} - \boldsymbol{r}'). \tag{5.26}$$

For the linear vibronic coupling,

$$\langle \boldsymbol{r}| \left(\frac{\partial \hat{H}}{\partial Q_\alpha}\right)_{\boldsymbol{R}^0} |\boldsymbol{r}'\rangle = \left(\frac{\partial \hat{H}}{\partial Q_\alpha}\right)_{\boldsymbol{R}^0} (\boldsymbol{r}, \boldsymbol{r}')\delta(\boldsymbol{r} - \boldsymbol{r}'). \tag{5.27}$$

In this book, we employ the following notation for a ket which depends on the coordinates,

$$|\Psi_n(\boldsymbol{r}, \boldsymbol{R}^0)\rangle := \int d_{\boldsymbol{r}} |\boldsymbol{r}\rangle\langle \boldsymbol{r}|\Psi_n\rangle = \int d_{\boldsymbol{r}} |\boldsymbol{r}\rangle \Psi_n(\boldsymbol{r}, \boldsymbol{R}^0). \tag{5.28}$$

In other words, we implicitly used $\boldsymbol{r}$-representation in this book. For an operator $\hat{A}$,

$$\hat{A} = \int dv_{\boldsymbol{r}'} \int dv_{\boldsymbol{r}''} |\boldsymbol{r}'\rangle\langle \boldsymbol{r}'|\hat{A}|\boldsymbol{r}''\rangle\langle \boldsymbol{r}''|. \tag{5.29}$$

For example,

$$V_{mn,\alpha} = \langle \Psi_m(\boldsymbol{r}, \boldsymbol{R}^0)| \left(\frac{\partial \hat{H}}{\partial Q_\alpha}\right)_{\boldsymbol{R}^0} |\Psi_n(\boldsymbol{r}, \boldsymbol{R}^0)\rangle \tag{5.30}$$

reads

$$\begin{aligned} V_{mn,\alpha} &= \int d_{\boldsymbol{r}} \Psi_m^*(\boldsymbol{r}, \boldsymbol{R}^0)\langle \boldsymbol{r}| \int dv_{\boldsymbol{r}'} \int dv_{\boldsymbol{r}''} |\boldsymbol{r}'\rangle\langle \boldsymbol{r}'| \left(\frac{\partial \hat{H}}{\partial Q_\alpha}\right)_{\boldsymbol{R}^0} \\ &\quad |\boldsymbol{r}''\rangle\langle \boldsymbol{r}''| \int d_{\boldsymbol{r}'''} |\boldsymbol{r}'''\rangle \Psi_n(\boldsymbol{r}''', \boldsymbol{R}^0) \\ &= \int d_{\boldsymbol{r}} dv_{\boldsymbol{r}'} dv_{\boldsymbol{r}''} d_{\boldsymbol{r}'''} \Psi_m^*(\boldsymbol{r}, \boldsymbol{R}^0)\delta(\boldsymbol{r} - \boldsymbol{r}') \left(\frac{\partial \hat{H}}{\partial Q_\alpha}\right)_{\boldsymbol{R}^0} \\ &\quad (\boldsymbol{r}', \boldsymbol{r}'')\delta(\boldsymbol{r}' - \boldsymbol{r}'')\delta(\boldsymbol{r}'' - \boldsymbol{r}''')\Psi_n(\boldsymbol{r}''', \boldsymbol{R}^0) \\ &= \int dv_{\boldsymbol{r}} \Psi_m^*(\boldsymbol{r}, \boldsymbol{R}^0) \left(\frac{\partial \hat{H}}{\partial Q_\alpha}\right)_{\boldsymbol{R}^0} (\boldsymbol{r}, \boldsymbol{r})\Psi_n(\boldsymbol{r}, \boldsymbol{R}^0). \end{aligned} \tag{5.31}$$

5.2 Group and Symmetry

5.2.1 Symmetry Operation and Wave Function: One-Dimensional Wall as an Example

Representation theory of group is useful in chemistry. Readers should learn the exact definitions and proofs in Ref. [1]. In this section, we will give some examples to show the way of applications of the concepts.

In quantum mechanics, it is required to evaluate a matrix element of an operator $\hat{O}$: $\langle \Psi(\boldsymbol{x}) | \hat{O} | \Phi(\boldsymbol{x}) \rangle$. This is the calculation of such an integral

$$\int d^3\boldsymbol{x} f(\boldsymbol{x}) = \int_{-\infty}^{+\infty} dx \int_{-\infty}^{+\infty} dy \int_{-\infty}^{+\infty} dz f(x, y, z), \tag{5.32}$$

where $f(\boldsymbol{x}) = \Psi(\boldsymbol{x}) \hat{O} \Phi(\boldsymbol{x})$. We can say that it is zero without a detailed calculation applying symmetry consideration.

For simplicity, we discuss a one-dimensional problem. Let us consider a real function $f(x)$ on $\mathbb{R}$, which is a set of real numbers. If $f(-x) = f(x)$ for any $x \in \mathbb{R}$, $f(x)$ is called an even function. If $f(-x) = -f(x)$ for any $x \in \mathbb{R}$, $f(x)$ is called an odd function. For an odd function $f(x)$,

$$\int_{-\infty}^{\infty} f(x)dx = 0. \tag{5.33}$$

This signifies that the symmetry of a function can tell us if the integral is necessarily zero since the independent variable replacement $x \to -x$ is the reflection which is a symmetry operation. For an even function $f(x)$ which is invariant under the symmetry operation, the integral can be finite.

When we define a symmetry operator $\hat{\sigma}$ acting on position x as

$$\hat{\sigma} : x \mapsto x' = -x, \tag{5.34}$$

the action on an even/odd function $f(x)$ is written as

$$\hat{\sigma} f(x) = (\hat{\sigma} f)(x) := f(-x) = \begin{cases} f(x) & \text{(even function)} \\ -f(x) & \text{(odd function)} \end{cases} \tag{5.35}$$

This signifies that an even/odd function $f(x)$ is the eigenfunction of $\hat{\sigma}$ belonging to the eigenvalue $+1$ or -1, respectively. For even/odd function f_Γ, $\Gamma = \mathrm{S}/\mathrm{AS}$, this can be rewritten as

$$(\hat{\sigma} f_\Gamma)(x) = D^\Gamma(\hat{\sigma}) f_\Gamma(x), \tag{5.36}$$

where

$$D^\Gamma(\hat{\sigma}) = \begin{cases} +1 & (\Gamma = \mathrm{S}) \\ -1 & (\Gamma = \mathrm{AS}) \end{cases}. \tag{5.37}$$

$D^\Gamma(\hat{\sigma})$ is called a representation of $\hat{\sigma}$. For two functions, f_{Γ_1} and g_{Γ_2}, the representation of the product $f_{\Gamma_1} g_{\Gamma_2}$ is obtained as

$$D^{\Gamma_1} D^{\Gamma_2} = \begin{cases} +1, & (\Gamma_1 = S, \Gamma_2 = S \;\text{ or }\; \Gamma_1 = AS, \Gamma_2 = AS) \\ -1, & (\Gamma_1 = S, \Gamma_2 = AS \;\text{ or }\; \Gamma_1 = AS, \Gamma_2 = S) \end{cases}. \tag{5.38}$$

As an example, for the one-dimensional wall, the Schrödinger equation is

$$\hat{H}\psi_n(x) = \left[-\frac{\hbar^2}{2m}\frac{d^2}{dx^2} + V(x)\right]\psi_n(x) = E_n\psi_n(x), \tag{5.39}$$

where

$$V(x) = \begin{cases} 0, & (-L/2 \le x \le L/2) \\ \infty, & (x < -L/2,\, L < x/2) \end{cases}. \tag{5.40}$$

The wave functions and eigenenergies are

$$\psi_n(x) = \begin{cases} \sqrt{\frac{2}{L}}\sin(\frac{n\pi}{L}x), & (n:\text{odd})\ (-L/2 \le x \le L/2) \\ \sqrt{\frac{2}{L}}\cos(\frac{n\pi}{L}x), & (n:\text{even})\ (-L/2 \le x \le L/2) \\ 0, & (x < -L/2,\, L < x/2) \end{cases} \tag{5.41}$$

and

$$E_n = \frac{\pi^2\hbar^2}{2mL^2}n^2. \tag{5.42}$$

Therefore, the eigenfunction ψ_n is either symmetric (S) (n: even) or anti-symmetric (AS) (n: odd). For an electric dipole transition from state m to n, the transition probability depends on the following integral:

$$\int_{-\infty}^{\infty} \psi_m^*(x)x\psi_n(x) \tag{5.43}$$

Note that $f(x) = x$ is transformed as AS under $\hat{S}$. We can easily see, for the product of three functions, $\psi_m^*(x)f(x)\psi_n(x)$, the integrand of the above integrand is transformed as S if the representations of $\psi_m^*(x)$ and $\psi_n(x)$ are transformed differently:

$$D^m D^f D^n = +1, \tag{5.44}$$

where D^m , D^f, and D^n are the representations of $\psi_m^*(x)$, $f(x)$, and $\psi_n(x)$, respectively. Therefore, if m is odd and n is even, or m is even and n is odd, the integrand is S, we can expect an optical transition between them. This is one of the selection rules.

5.2.2 *Group and Its Representation: Two-Dimensional Wall as an Example*

In the previous section, the representation is a scalar. For more than a two-dimensional case, a representation can become matrix, $\mathbf{D}$.

Before we proceed with the two-dimensional case, some definitions discussed in the previous section are extended for a three-dimensional case. For $\boldsymbol{x} \in \mathbb{R}^3$, a symmetry operator $\hat{R}$ such as rotation about n-fold symmetry axis C_n, reflection for a plane σ, and inversion i,

$$\hat{R} : \boldsymbol{x} = \begin{pmatrix} x \\ y \\ z \end{pmatrix} \mapsto \boldsymbol{x}' = \begin{pmatrix} x' \\ y' \\ z' \end{pmatrix} = \mathbf{R}\boldsymbol{x}, \tag{5.45}$$

where $\mathbf{R}$ is the matrix representation of $\hat{R}$. In general, a symmetry operation $\hat{R}$ acting on a function f defined in $\mathbb{R}^3$ operates as follows:

$$\hat{R} : f(\boldsymbol{x}) \mapsto f'(x) = (\hat{R}f)(\boldsymbol{x}) = f(\mathbf{R}^{-1}\boldsymbol{x}). \tag{5.46}$$

We will discuss the two-dimensional wall as an example.

$$\hat{H}\psi_{n_x,n_y}(x, y) = \left[-\frac{\hbar^2}{2m}\left(\frac{\partial^2}{\partial x^2} + \frac{\partial^2}{\partial y^2}\right) + V(x, y)\right]\psi_{n_x,n_y}(x, y) = E_{n_x,n_y}\psi_{n_x,n_y}(x, y), \tag{5.47}$$

where

$$V(x, y) = \begin{cases} 0, & (-L_x/2 \le x \le L_x/2 \ \text{ and } \ -L_y/2 \le y \le L_y/2) \\ \infty, & (\text{otherwise}) \end{cases}. \tag{5.48}$$

The Hamiltonian is invariant under the following symmetry operations:

$$\sigma_x = \begin{pmatrix} -1 & 0 \\ 0 & 1 \end{pmatrix}, \quad \sigma_y = \begin{pmatrix} 1 & 0 \\ 0 & -1 \end{pmatrix}. \tag{5.49}$$

The wave functions inside the wall and eigenenergies are (Table 5.1)

$$\psi_{n_x,n_y}(x, y) = \begin{cases} \sqrt{\frac{2}{L_x}}\sin(\frac{n_x\pi}{L_x}x)\sqrt{\frac{2}{L_y}}\sin(\frac{n_y\pi}{L_y}y) = \psi^{\text{AS,AS}}, & (n_x : \text{odd}, n_y : \text{odd}) \\ \sqrt{\frac{2}{L_x}}\sin(\frac{n_x\pi}{L_x}x)\sqrt{\frac{2}{L_y}}\cos(\frac{n_y\pi}{L_y}y) = \psi^{\text{AS,S}}, & (n_x : \text{odd}, n_y : \text{even}) \\ \sqrt{\frac{2}{L_x}}\cos(\frac{n_x\pi}{L_x}x)\sqrt{\frac{2}{L_y}}\sin(\frac{n_y\pi}{L_y}y) = \psi^{\text{S,AS}}, & (n_x : \text{even}, n_y : \text{odd}) \\ \sqrt{\frac{2}{L_x}}\cos(\frac{n_x\pi}{L_x}x)\sqrt{\frac{2}{L_y}}\cos(\frac{n_y\pi}{L_y}y) = \psi^{\text{S,S}}, & (n_x : \text{even}, n_y : \text{even}) \end{cases} \tag{5.50}$$

and

$$E_{n_x,n_y} = \frac{\pi^2\hbar^2}{2mL_x^2}n_x^2 + \frac{\pi^2\hbar^2}{2mL_y^2}n_y^2. \tag{5.51}$$

Further condition, $L_x = L_y = L$, induces a higher symmetry in the system. In addition to the symmetry operations mentioned above, the Hamiltonian is invariant under the following symmetry operation:

$$\mathbf{C}_4 = \begin{pmatrix} 0 & -1 \\ 1 & 0 \end{pmatrix}. \tag{5.52}$$

Table 5.1 Transformation of the wave functions

	$\hat{E}$	$\hat{\sigma}_x$	$\hat{\sigma}_y$
$\psi^{\mathrm{S,S}}$	1	1	1
$\psi^{\mathrm{S,AS}}$	1	1	–1
$\psi^{\mathrm{AS,S}}$	1	–1	1
$\psi^{\mathrm{AS,AS}}$	1	–1	–1

The solutions are

$$\psi_{n_x,n_y}(x,y)=\begin{cases}\frac{2}{L}\sin(\frac{n_x\pi}{L}x)\sin(\frac{n_y\pi}{L}y) & (n_x:\text{odd}, n_y:\text{odd})\\ \frac{2}{L}\sin(\frac{n_x\pi}{L}x)\cos(\frac{n_y\pi}{L}y) & (n_x:\text{odd}, n_y:\text{even})\\ \frac{2}{L}\cos(\frac{n_x\pi}{L}x)\sin(\frac{n_y\pi}{L}y) & (n_x:\text{even}, n_y:\text{odd})\\ \frac{2}{L}\cos(\frac{n_x\pi}{L}x)\cos(\frac{n_y\pi}{L}y) & (n_x:\text{even}, n_y:\text{even})\end{cases} \tag{5.53}$$

and

$$E_{n_x,n_y}=\frac{\pi^2\hbar^2}{2mL^2}\left(n_x^2+n_y^2\right). \tag{5.54}$$

For $(n_x,n_y)=(k,l)$ and $(n_x,n_y)=(l,k)$ with $k\neq l$, the energy level E_{n_x,n_y} is doubly degenerate:

$$E_{n_x,n_y}=\frac{\pi^2\hbar^2}{2mL^2}\left(k^2+l^2\right). \tag{5.55}$$

$$\mathbf{C}_4^{-1}:\boldsymbol{x}\mapsto\boldsymbol{x}'=\begin{pmatrix}0&1\\-1&0\end{pmatrix}\begin{pmatrix}x\\y\end{pmatrix}=\begin{pmatrix}-y\\x\end{pmatrix}. \tag{5.56}$$

For $n_x=2m+1,\ n_y=2l$,

$$\psi_{2m+1,2l}(x,y)=\frac{2}{L}\sin\left(\frac{(2m+1)\pi}{L}x\right)\cos\left(\frac{2l\pi}{L}y\right), \tag{5.57}$$

and for $n_x=2l,\ n_y=2m+1$

$$\psi_{2l,2m+1}(x,y)=\frac{2}{L}\cos\left(\frac{2l\pi}{L}x\right)\sin\left(\frac{(2m+1)\pi}{L}y\right). \tag{5.58}$$

The two functions transformed each other under $\hat{C}_4$:

$$(\hat{C}_4\psi_{2m+1,2l})(x,y)=\psi_{2m+1,2l}(y,-x)=\psi_{2l,2m+1}(x,y), \tag{5.59}$$

$$(\hat{C}_4\psi_{2l,2m+1})(x,y)=\psi_{2l,2m+1}(y,-x)=-\psi_{2m+1,2l}(x,y). \tag{5.60}$$

Therefore, the representation of $\hat{C}_4$, $\mathbf{D}(\hat{C}_4)$, is expressed in terms of $(\psi_{2m+1,2l}, \psi_{2l,2m+1})$ by the following 2×2 matrix:

$$\hat{C}_4(\psi_{2m+1,2l}, \psi_{2l,2m+1}) = (\psi_{2m+1,2l}, \psi_{2l,2m+1})\mathbf{D}(\hat{C}_4), \tag{5.61}$$

where

$$\mathbf{D}^E(\hat{C}_4) = \begin{pmatrix} 0 & -1 \\ 1 & 0 \end{pmatrix}. \tag{5.62}$$

The three symmetry operations, $\hat{C}_4$, $\hat{\sigma}_x$, and $\hat{\sigma}_y$, generate the C_{4v} point group. The representation matrix for $(\psi_{2m+1,2l}, \psi_{2l,2m+1})$ is tabulated in Table 5.2. This representation is called E representation.

Since there occurs a product of wave functions in a matrix element $\langle \Psi_m | \hat{O} | \Psi_n \rangle$, the transformation property of the product $\Psi_m \Psi_n$ is useful to evaluate it. We can construct other basis set by making products between $\psi_{2m+1,2l}$ and $\psi_{2l,2m+1}$:

$$\begin{aligned}
&\psi_{2m+1,2l}(x_1, y_1)\psi_{2m+1,2l}(x_2, y_2),\\
&\psi_{2m+1,2l}(x_1, y_1)\psi_{2l,2m+1}(x_2, y_2),\\
&\psi_{2l,2m+1}(x_1, y_1)\psi_{2m+1,2l}(x_2, y_2),\\
&\psi_{2l,2m+1}(x_1, y_1)\psi_{2l,2m+1}(x_2, y_2).
\end{aligned} \tag{5.63}$$

This product representation $E \times E$ is obtained as follows:

$$\mathbf{D}^{E \times E}(\hat{C}_4) = \begin{pmatrix} 0 & 0 & 0 & 1 \\ 0 & 0 & -1 & 0 \\ 0 & -1 & 0 & 0 \\ 1 & 0 & 0 & 0 \end{pmatrix}, \tag{5.64}$$

$$\mathbf{D}^{E \times E}(\hat{\sigma}_x) = \begin{pmatrix} 1 & 0 & 0 & 0 \\ 0 & -1 & 0 & 0 \\ 0 & 0 & -1 & 0 \\ 0 & 0 & 0 & 1 \end{pmatrix}, \tag{5.65}$$

and

$$\mathbf{D}^{E \times E}(\hat{\sigma}_y) = \begin{pmatrix} 1 & 0 & 0 & 0 \\ 0 & -1 & 0 & 0 \\ 0 & 0 & -1 & 0 \\ 0 & 0 & 0 & 1 \end{pmatrix}. \tag{5.66}$$

Applying the following basis transformation:

Table 5.2 Irreducible representation matrices of C_{4v}

$\hat{R}$	$\hat{E}$	$\hat{C}_4, \hat{C}_4^3$	$\hat{C}_2$	$2\hat{\sigma}_v$	$2\hat{\sigma}_d$	basis
$\mathbf{D}^E(\hat{R})$	$\begin{pmatrix} 1 & 0 \\ 0 & 1 \end{pmatrix}$	$\begin{pmatrix} 0 & -1 \\ 1 & 0 \end{pmatrix}, \begin{pmatrix} 0 & -1 \\ 1 & 0 \end{pmatrix}$	$\begin{pmatrix} -1 & 0 \\ 0 & -1 \end{pmatrix}$	$\begin{pmatrix} -1 & 0 \\ 0 & 1 \end{pmatrix}, \begin{pmatrix} 1 & 0 \\ 0 & -1 \end{pmatrix}$	$\begin{pmatrix} 0 & 1 \\ 1 & 0 \end{pmatrix}, \begin{pmatrix} 0 & -1 \\ -1 & 0 \end{pmatrix}$	$(\psi_{2m+1,2l}, \psi_{2l,2m+1})$
$\mathbf{D}^{A_1}(\hat{R})$	(1)	(1)	(1)	(1)	(1)	
$\mathbf{D}^{A_2}(\hat{R})$	(1)	(1)	(1)	(−1)	(−1)	
$\mathbf{D}^{B_1}(\hat{R})$	(1)	(−1)	(1)	(1)	(−1)	
$\mathbf{D}^{B_2}(\hat{R})$	(1)	(−1)	(1)	(−1)	(1)	

$$\psi_{A_1} = \frac{1}{\sqrt{2}}\psi_{2m+1,2l}(x_1, y_1)\psi_{2m+1,2l}(x_2, y_2) + \frac{1}{\sqrt{2}}\psi_{2l,2m+1}(x_1, y_1)\psi_{2l,2m+1}(x_2, y_2), \tag{5.67}$$

$$\psi_{B_1} = -\frac{1}{\sqrt{2}}\psi_{2m+1,2l}(x_1, y_1)\psi_{2m+1,2l}(x_2, y_2) + \frac{1}{\sqrt{2}}\psi_{2l,2m+1}(x_1, y_1)\psi_{2l,2m+1}(x_2, y_2), \tag{5.68}$$

$$\psi_{B_2} = \frac{1}{\sqrt{2}}\psi_{2m+1,2l}(x_1, y_1)\psi_{2l,2m+1}(x_2, y_2) + \frac{1}{\sqrt{2}}\psi_{2l,2m+1}(x_1, y_1)\psi_{2m+1,2l}(x_2, y_2), \tag{5.69}$$

$$\psi_{A_2} = \frac{1}{\sqrt{2}}\psi_{2m+1,2l}(x_1, y_1)\psi_{2l,2m+1}(x_2, y_2) - \frac{1}{\sqrt{2}}\psi_{2l,2m+1}(x_1, y_1)\psi_{2m+1,2l}(x_2, y_2), \tag{5.70}$$

the representation matrices $\mathbf{D}(\hat{R})$ are block diagonalized:

$$\mathbf{UD}^{E\times E}(\hat{C}_4)\mathbf{U}^{-1} = \begin{pmatrix} 1 & 0 & 0 & 0 \\ 0 & -1 & 0 & 0 \\ 0 & 0 & -1 & 0 \\ 0 & 0 & 0 & 1 \end{pmatrix}, \quad \mathbf{UD}^{E\times E}(\hat{C}_2)\mathbf{U}^{-1} = \begin{pmatrix} 1 & 0 & 0 & 0 \\ 0 & 1 & 0 & 0 \\ 0 & 0 & 1 & 0 \\ 0 & 0 & 0 & 1 \end{pmatrix}, \tag{5.71}$$

$$\mathbf{UD}^{E\times E}(\hat{\sigma}_x)\mathbf{U}^{-1} = \begin{pmatrix} 1 & 0 & 0 & 0 \\ 0 & 1 & 0 & 0 \\ 0 & 0 & -1 & 0 \\ 0 & 0 & 0 & -1 \end{pmatrix}, \quad \mathbf{UD}^{E\times E}(\hat{C}_4\hat{\sigma}_y)\mathbf{U}^{-1} = \begin{pmatrix} 1 & 0 & 0 & 0 \\ 0 & -1 & 0 & 0 \\ 0 & 0 & 1 & 0 \\ 0 & 0 & 0 & -1 \end{pmatrix}, \tag{5.72}$$

where

$$\mathbf{U} = \frac{1}{\sqrt{2}}\begin{pmatrix} 1 & 0 & 0 & 1 \\ -1 & 0 & 0 & 1 \\ 0 & 1 & 1 & 0 \\ 0 & -1 & 1 & 0 \end{pmatrix}. \tag{5.73}$$

These are summarized as follows:

$$\mathbf{UD}^{E\times E}(\hat{R})\mathbf{U}^{-1} = \begin{pmatrix} \mathbf{D}^{A_1}(\hat{R}) & 0 & 0 & 0 \\ 0 & \mathbf{D}^{B_1}(\hat{R}) & 0 & 0 \\ 0 & 0 & \mathbf{D}^{B_2}(\hat{R}) & 0 \\ 0 & 0 & 0 & \mathbf{D}^{A_2}(\hat{R}) \end{pmatrix}. \tag{5.74}$$

Table 5.3 Characters of the irreducible representations of C_{4v}

$\hat{R}$	$\hat{E}$	$2\hat{C}_4$	$\hat{C}_2$	$2\hat{\sigma}_v$	$2\hat{\sigma}_d$		
A_1	1	1	1	1	1	z	$x^2 + y^2, z^2$
A_2	1	1	1	-1	-1		
B_1	1	-1	1	1	-1		$x^2 - y^2$
B_1	1	-1	1	-1	1		xy
E	2	0	-2	0	0	(x, y)	(xz, yz)

Table 5.4 Characters of $D_{\infty h}$

$D_{\infty h}$	E	$2C_\infty^\phi$	$\cdots$	$\infty\sigma_v$	i	$2S_\infty^\phi$	$\cdots$	∞C_2		
$\Sigma_g^+(A_{1g})$	1	1	$\cdots$	1	1	1	$\cdots$	1		$x^2 + y^2, z^2$
$\Sigma_g^-(A_{2g})$	1	1	$\cdots$	–1	1	1	$\cdots$	–1	R_z	
$\Pi_g(E_g)$	2	$2\cos\phi$	$\cdots$	0	2	$-2\cos\phi$	$\cdots$	0	(R_x, R_y)	(xz, yz)
$\Sigma_u^+(A_{1u})$	1	1	$\cdots$	1	–1	–1	$\cdots$	–1	z	
$\Sigma_u^-(A_{2u})$	1	1	$\cdots$	–1	–1	–1	$\cdots$	1		
$\Pi_u(E_u)$	2	$2\cos\phi$	$\cdots$	0	–2	$2\cos\phi$	$\cdots$	0	(x, y)	

This result signifies that the product representation is reduced to the direct sum of A_1, A_2, B_1, and B_2 representations. These elementary representations A_1, A_2, B_1, B_2, and E are called irreducible representations (irrep) . In Eq. (5.70), if we make the following transformation, $x_1 \leftrightarrow x_2$, $y_1 \leftrightarrow y_2$, ψ_{A_1}, ψ_{B_1}, and ψ_{B_2} are invariant. The representation in terms of such a representation is called a symmetric product representation. On the other hand, since ψ_{A_1} is anti-symmetric, its representation is called an anti-symmetric product representation. In Tables 5.9, 5.10, and 5.11, symmetric product representations are denoted by [ ], and anti-symmetric representations are { }.

In general, a wave function belongs to one of the irreps of the group under which the Hamiltonian is invariant. To evaluate a matrix element, $\langle \Psi_m | \hat{O} | \Psi_n \rangle$, we can derive the condition for a non-zero integral:

$$\Gamma_m \times \Gamma_n \quad \text{contains} \quad \Gamma_{\hat{O}}, \tag{5.75}$$

where Γ_m, Γ_n, and $\Gamma_{\hat{O}}$ are the irreps of Ψ_m, Ψ_n, and $\hat{O}$, respectively. This is because the product representation between itself, $\Gamma \times \Gamma$, always contains a totally symmetric representation A_1, and the integrand contains a non-vanishing component after the integration. Accordingly, it is necessary to decompose the product representation $\Gamma_m \times \Gamma_n$ into the direct sum of irreps. This can be easily done by using a character $\chi(\Gamma)$ defined by the trace of a representation matrix. The character table of C_{4v} is shown in Table 5.3. The characters of the other groups which appeared in this book are tabulated in Tables 5.4, 5.5, 5.6, 5.7, and 5.8.

Table 5.5 Characters of C_{2v}

C_{2v}	E	C_2	$\sigma_v(xz)$	$\sigma_v'(yz)$		
A_1	1	1	1	1	z	x^2, y^2, z^2
A_2	1	1	–1	–1	R_z	xy
B_1	1	–1	1	–1	x, R_y	xz
B_2	1	–1	–1	1	y, R_x	yz

Table 5.6 Characters of C_{3v}

C_{3v}	E	$2C_3$	$3\sigma_v$		
A_1	1	1	1	z	$x^2 + y^2, z^2$
A_2	1	1	–1	R_z	
E	2	–1	0	(x, y) (R_x, R_y)	($x^2 - y^2$, 2xy)(xz, yz)

Table 5.7 Characters of D_{3h}

D_{3h}	E	$2C_3$	$3C_2$	σ_h	$2S_3$	$3\sigma_v$		
A_1'	1	1	1	1	1	1		$x^2 + y^2, z^2$
A_2'	1	1	–1	1	1	–1	R_z	
E'	2	–1	0	2	–1	0	(x, y)	$(x^2 - y^2, 2xy)$
A_1''	1	1	1	–1	–1	–1		
A_2''	1	1	–1	–1	–1	1	z	
E''	2	–1	0	–2	1	0	(R_x, R_y)	(xy, yz)

Table 5.8 Characters of D_{6h}

D_{6h}	E	$2C_6$	$2{C_6}^2$	C_2	$3C_2'$	$3C_2''$	i	$2S_3$	$2S_6$	σ_h	$3\sigma_d$	$3\sigma_v$		
A_{1g}	1	1	1	1	1	1	1	1	1	1	1	1		$x^2 + y^2, z^2$
A_{1u}	1	1	1	1	1	1	–1	–1	–1	–1	–1	–1		
A_{2g}	1	1	1	1	–1	–1	1	1	1	1	–1	–1	R_z	
A_{2u}	1	1	1	1	–1	–1	–1	–1	–1	–1	1	1	z	
B_{1g}	1	–1	1	–1	1	–1	1	–1	1	–1	1	–1		
B_{1u}	1	–1	1	–1	1	–1	–1	1	—1	1	–1	1		
B_{2g}	1	–1	1	–1	–1	1	1	–1	1	–1	–1	1		
B_{2u}	1	–1	1	–1	–1	1	–1	1	–1	1	1	–1		
E_{1g}	2	1	–1	–2	0	0	2	1	–1	–2	0	0	(R_x, R_y)	(xy, yz)
E_{1u}	2	1	–1	–2	0	0	–2	–1	1	2	0	0	(x, y)	
E_{2g}	2	–1	–1	2	0	0	2	–1	–1	2	0	0		$(x^2 - y^2, xy)$
E_{2u}	2	–1	–1	2	0	0	–2	1	1	–2	0	0		

Table 5.9 Direct products of $D_{\infty h}$

$D_{\infty h}$	A_{1g}	A_{2g}	E_g	A_{1u}	A_{2u}	E_u
A_{1g}	A_{1g}	A_{2g}	E_g	A_{1u}	A_{2u}	E_u
A_{2g}		A_{1g}	E_g	A_{2u}	A_{1u}	E_u
E_g			$[A_{1g} + E_g] + \{A_{2g}\}$	E_g	E_g	$A_{1u} + A_2 + E_u$
A_{1u}				A_{1g}	A_{2g}	E_g
A_{2u}					A_{1g}	E_g
E_u						$[A_{1g} + E_g] + \{A_{2g}\}$

Table 5.10 Direct products of D_{3h}

D_{3h}	A_1'	A_2'	E'	A_1''	A_2''	E''
A_1'	A_1'	A_2'	E'	A_1''	A_2''	E''
A_2'		A_1'	E'	A_2''	A_1''	E''
E'			$[A_1' + E'] + \{A_2'\}$	E''	E'	$A_1'' + A_2'' + E''$
A_1''				A_1'	A_2'	E'
A_2''					A_1'	E'
E''						$[A_1' + E'] + \{A_2'\}$

Table 5.11 Direct products of D_{6h}

D_{6h}	A_{1g}	A_{1u}	A_{2g}	A_{2u}	B_{1g}	B_{1u}	B_{2g}	B_{2u}	E_{1g}	E_{1u}	E_{2g}	E_{2u}
A_{1g}	A_{1g}	A_{1u}	A_{2g}	A_{2u}	B_{1g}	B_{1u}	B_{2g}	B_{2u}	E_{1g}	E_{1u}	E_{2g}	E_{2u}
A_{1u}		A_{1g}	A_{2u}	A_{2g}	B_{1u}	B_{1g}	B_{2u}	B_{2g}	E_{1u}	E_{1g}	E_{2u}	E_{2g}
A_{2g}			A_{1g}	A_{1u}	B_{2g}	B_{2u}	B_{1g}	B_{1u}	E_{1g}	E_{1u}	E_{2g}	E_{2u}
A_{2u}				A_{1g}	B_{2u}	B_{2g}	B_{1u}	B_{1g}	E_{1u}	E_{1g}	E_{2u}	E_{2g}
B_{1g}					A_{1g}	A_{1u}	A_{2g}	A_{2u}	E_{2g}	E_{2u}	E_{1g}	E_{1u}
B_{1u}						A_{1g}	A_{2u}	A_{2g}	E_{2u}	E_{2g}	E_{1u}	E_{1g}
B_{2g}					–	–	A_{1g}	A_{1u}	E_{2g}	E_{2u}	E_{1g}	E_{1u}
B_{2u}								A_{1g}	E_{2u}	E_{2g}	E_{1u}	E_{1g}
E_{1g}									$([A_{1g}+E_{2g}]+\{A_{2g}\})$	$(A_{1u}+A_{2u}+E_{2u})$	$(B_{1g}+B_{2g}+E_{1g})$	$(B_{1u}+B_{2u}+E_{1u})$
E_{1u}										$([A_{1g}+E_{2g}]+\{A_{2g}\})$	$(B_{1u}+B_{2u}+E_{1u})$	$(B_{1g}+B_{2g}+E_{1g})$
E_{2g}											$([A_{1g}+E_{2g}]+\{A_{2g}\})$	$(A_{1u}+A_{2u}+E_{2u})$
E_{2u}												$([A_{1g}+E_{2g}]+\{A_{2g}\})$

From Table 5.3, we can derive the characters of the product representation $\Gamma_E \times \Gamma_E$ and its decomposition into the direct sum of the irreps:

$\hat{R}$	$\hat{E}$	$2\hat{C}_4$	$\hat{C}_2$	$2\hat{\sigma}_v$	$2\hat{\sigma}_d$
$E \times E$	$2 \times 2 = 4$	$0 \times 0 = 0$	$(-2) \times (-2) = 4$	$0 \times 0 = 0$	$0 \times 0 = 0$
$A_1 + A_2 + B_1 + B_4$	$1+1+1+1 = 4$	$1+1-1-1 = 0$	$1+1+1+1 = 4$	$1-1+1-1 = 0$	$1-1-1+1 = 0$

Accordingly,

$$E \times E = [A_1 + B_1 + B_2] + \{A_2\}. \tag{5.76}$$

The decompositions of the product representations in the other groups which appeared in this book are summarized in Tables 5.9, 5.10, and 5.11. Since the z-component of the electric dipole operator $\hat{\mu}_z = -ez$ belongs to A_1, the z-component of the transition dipole moment between the states which belong to E irrep is non-zero:

$$\Gamma_{\text{SOMO}} \times \Gamma_{\text{SOMO}} = E \times E = A_1 + B_1 + B_2 \ni A_1 = \Gamma_z. \tag{5.77}$$

Note that the anti-symmetric product is vanishing because $x_1 = x_2,\ y_1 = y_2$ in the integrand of Eq. (5.32).

5.2.3 *Point Group and Its Irreducible Representation: Triatomic Molecules as an Example*

In this section, we discuss the use of the symmetry of triatomic molecules, NH_2, H_2O, and A_3. Figure 1.2b shows the molecular orbitals of H_2A with the linear structures, the point group of which is $D_{\infty h}$. The SOMO level of NH_2 with the linear geometry is doubly degenerate (E irrep). As shown in Fig. 1.2b, the SOMO is anti-symmetric (u irrep) for the inversion. Therefore, from Table 5.4, the SOMOs belong to $E_u(\Pi_u)$ irrep, and are denoted by the π_u-orbitals. The ground electronic state of the linear NH_2 is $^2\Pi_u$ or ^2E_u.

Figure 5.1 shows the normal modes of a linear H_2A. There are four modes, which belong to A_{1g}, A_{1u}, and doubly degenerate E_u irreps:

$$\Gamma_{\text{vib}} = A_{1g} + A_{1u} + E_u. \tag{5.78}$$

The selection rule of the linear vibronic coupling between electronic states Ψ_m and Ψ_n is

$$\Gamma_m \times \Gamma_\nu \times \Gamma_n \quad \text{contains} \quad A_{1g}, \tag{5.79}$$

where Γ_m and Γ_n denote the irreps of the electronic states, and Γ_ν is the irrep of the normal mode which couples with the electronic states. From Table 5.9, the product $E_u \times E_u$ is decomposed as

Fig. 5.1 Vibrational normal modes of $H_2A(D_{\infty h})$

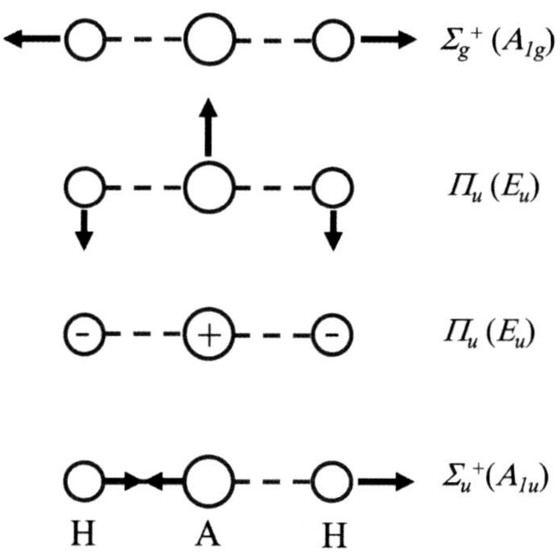

$$E_u \times E_u = [A_{1g} + E_g] + \{A_{2g}\}, \tag{5.80}$$

or

$$[E_u^2] = A_{1g} + E_g, \quad \{E_u^2\} = A_{2g}. \tag{5.81}$$

Thus, the active mode of the linear coupling is only the A_{1g} mode. It should be noted that, as shown in (5.78), there is no E_g mode in the vibrational modes in H_2A, and the JT distortion, which is due to the linear coupling, is impossible. On the other hand, a quadratic coupling of mode E_u is possible since the product of Q_{E_u}s contains E_g: from Table 5.9,

$$E_u \times E_u = [A_{1g} + E_g] + \{A_{2g}\}. \tag{5.82}$$

Therefore, the quadratic vibronic coupling is not vanishing and gives rise to Renner–Teller effect.

5.3 Molecular Vibrations and Normal Modes

Let us consider the nuclear motion of an N-atom molecule. Its displacement vector $\Delta \boldsymbol{R}$ from the equilibrium nuclear configuration $\boldsymbol{R}^0$ is written by

$$\Delta \boldsymbol{R} := \begin{pmatrix} \vdots \\ \Delta X_{A\xi} \\ \vdots \end{pmatrix}, \tag{5.83}$$

where $\Delta X_{A\xi}$ denotes the ξ-component ($\xi = x, y, z$) of the displacement of an Ath nucleus in the Cartesian coordinate system. Within the second-order approximation, the nuclear Hamiltonian $\hat{H}$ is given by

$$\hat{H} := \sum_{A} \sum_{\xi=x,y,z} \left[-\frac{\hbar^2}{2m_A} \frac{\partial^2}{\partial X_{A\xi}^2} \right] + \frac{1}{2} \Delta \boldsymbol{R}^{\mathrm{T}} \mathbf{K} \Delta \boldsymbol{R}, \tag{5.84}$$

where m_A stands for the mass of an Ath nucleus, $\hbar$ represents the reduced Planck's constant (Dirac constant), and $\mathbf{K} := (K_{A\xi,B\xi'})$ is a Hessian, i.e., a real symmetric matrix consisting of the second-order partial derivatives of the potential energy $E(\boldsymbol{R})$,

$$K_{A\xi,B\xi'} := \left(\frac{\partial^2 E(\boldsymbol{R})}{\partial X_{A\xi} \partial X_{B\xi'}} \right)_{\boldsymbol{R}^0}. \tag{5.85}$$

We subsequently convert the real coordinate ΔX_{Ar} to the mass-weighted coordinate $\Delta X'_{Ar} := \sqrt{m_A} \Delta X_{A\xi}$:

$$\Delta \boldsymbol{R}' := \begin{pmatrix} \vdots \\ \Delta X'_{A\xi} \\ \vdots \end{pmatrix} = \begin{pmatrix} \vdots \\ \sqrt{m_A} \Delta X_{A\xi} \\ \vdots \end{pmatrix}. \tag{5.86}$$

Accordingly, $\Delta \boldsymbol{R}'$ is expressed as

$$\Delta \boldsymbol{R}' = \boldsymbol{M}^{-1} \Delta \boldsymbol{R}, \tag{5.87}$$

where $\boldsymbol{M}$ is a real diagonal matrix whose element $M_{A\xi,B\xi'}$ is

$$M_{A\xi,B\xi'} = \frac{1}{\sqrt{m_A}} \delta_{AB} \delta_{\xi\xi'}. \tag{5.88}$$

$\hat{H}$ can therefore be rewritten as

$$\hat{H} = \sum_{A} \sum_{\xi=x,y,z} \left[-\frac{\hbar^2}{2} \frac{\partial^2}{\partial \Delta X'^2_{A\xi}} \right] + \frac{1}{2} \Delta \boldsymbol{R}'^{\mathrm{T}} \mathbf{K}' \Delta \boldsymbol{R}', \tag{5.89}$$

where $\mathbf{K}' := (K'_{A\xi,B\xi'})$ is a mass-weighted Hessian whose element $K'_{A\xi,B\xi'}$ is

$$K'_{A\xi,B\xi'} := \frac{K_{A\xi,B\xi'}}{\sqrt{m_A m_B}}. \tag{5.90}$$

We here discuss the eigenvalue problem of the Hessian $\mathbf{K}'$:

$$\mathbf{K}' \boldsymbol{u}_\alpha = \omega_\alpha^2 \boldsymbol{u}_\alpha \quad (\alpha = 1, 2, \ldots, 3N). \tag{5.91}$$

Since $\mathbf{K}'$ is a real symmetric matrix, ω_α^2 is a real number, and $\boldsymbol{u}_\alpha$ $(\alpha = 1, 2, \ldots, 3N)$ can be chosen such that they are orthonormal with each other. ω_α's are called frequencies, and $\boldsymbol{u}_\alpha$'s are normal vectors related to translation, rotation, and vibration in the mass-weighted space.

Any mass-weighted displacement vector $\Delta \boldsymbol{R}'$ can be represented as a linear combination of $\boldsymbol{u}_\alpha$ $(\alpha = 1, 2, \ldots, 3N)$,

$$\Delta \boldsymbol{R}' = \sum_\alpha Q_\alpha \boldsymbol{u}_\alpha, \tag{5.92}$$

where Q_α's are called mass-weighted normal coordinates. Consequently, $\hat{H}$ is rewritten by

$$\hat{H} = \sum_\alpha \left[-\frac{\hbar^2}{2} \frac{\partial^2}{\partial Q_\alpha^2} + \frac{1}{2} \omega_\alpha^2 Q_\alpha^2 \right], \tag{5.93}$$

because of the orthonormality, $\boldsymbol{u}_\alpha^{\mathrm{T}} \boldsymbol{u}_\beta = \delta_{\alpha\beta}$.

From Eqs. 5.86 and 5.92, $\Delta \boldsymbol{R}$ is expressed as

$$\Delta \boldsymbol{R} = \boldsymbol{M} \Delta \boldsymbol{R}' = \sum_\alpha Q_\alpha \boldsymbol{M} \boldsymbol{u}_\alpha. \tag{5.94}$$

We herein introduce q_α and $\boldsymbol{v}_\alpha$ as

$$q_\alpha := |\boldsymbol{M} \boldsymbol{u}_\alpha| Q_\alpha, \tag{5.95}$$

$$\boldsymbol{v}_\alpha := \frac{\boldsymbol{M} \boldsymbol{u}_\alpha}{|\boldsymbol{M} \boldsymbol{u}_\alpha|}, \tag{5.96}$$

and thereby obtain the following expression:

$$\Delta \boldsymbol{R} = \sum_\alpha q_\alpha \boldsymbol{v}_\alpha, \tag{5.97}$$

where $\boldsymbol{v}_\alpha$'s are called normal vectors in the real space, and q_α's are called real normal coordinates. It should be noted that $\boldsymbol{v}_\alpha$'s are not orthogonal in general. Its exception is homonuclear molecules.

Finally, we define a reduced mass μ_α of mode α as

$$\mu_\alpha := \frac{1}{|\boldsymbol{M} \boldsymbol{u}_\alpha|^2}. \tag{5.98}$$

Therefore, $\hat{H}$ can be rewritten as

$$\hat{H} = \sum_\alpha \left[-\frac{\hbar^2}{2\mu_\alpha} \frac{\partial^2}{\partial q_\alpha^2} + \frac{1}{2} \mu_\alpha \omega_\alpha^2 q_\alpha^2 \right], \tag{5.99}$$

which means the motion of $3N$ virtual particles with masses μ_α's. Additionally, Eqs. 5.95 and 5.98 lead to the relation between q_α and Q_α:

$$q_\alpha = \frac{Q_\alpha}{\sqrt{\mu_\alpha}}. \quad (5.100)$$

5.4 Vibronic Coupling Density

The vibronic coupling density (VCD) and the vibronic coupling constant (VCC) can be explicitly defined by the *ab initio* molecular Hamiltonian. And they can give the quantitative evaluation of the force applied under the chemical deformation process. We give the definition and evaluation of VCD and VCC in this section.

5.4.1 *Vibronic Coupling Constant and Vibronic Coupling Density*

In most quantum chemistry calculations, the Schrödinger equation in which the kinetic energy operator of nuclei is neglected is solved assuming that the masses of the nuclei are heavier than that of an electron. However, the motions of nuclei sometimes play some important roles, as we describe in this monograph. An adiabatic approximation is usually employed to solve the Schrödinger equation, including the nuclear kinetic energy operator. As a result of the adiabatic approximation, a molecular wave function, or a vibronic wave function, is represented as the product of electronic and vibrational wave functions [2–4]. A vibronic wave function is expanded using electronic wave functions for a certain nuclear configuration in the crude adiabatic approximation. Thus, the dependencies of nuclear configurations are described only by nuclear wave functions. On the other hand, in the Born–Oppenheimer approximation, a vibronic wave function is expanded so that both electronic and vibrational wave functions depend on nuclear configurations.

We have investigated vibronic coupling, the interaction between vibrational and electronic motions, within the crude adiabatic approximation [5, 6]. Vibronic coupling is quantified by a vibronic coupling constant (VCC). Vibronic coupling density (VCD) is a density form of the VCC. Using the VCD, we can discuss the reason for the small or large values of VCC based on the electronic and vibrational structures. Moreover, VCD analysis enables us to control the VCC by a chemical modification.

A well-known phenomenon induced by vibronic coupling is the Jahn–Teller (JT) effect, in which vibronic coupling lowers the geometrical symmetry of a non-linear molecule with a degenerate electronic state to stabilize the system by a symmetry-lowering deformation to lift the degeneracy of the electronic state [7–9]. The VCC and VCD have been evaluated for C_3H_3 in Section 3.5, and the VCC and VCD have well-characterized the systems such as cyclopentadienyl [10, 11], benzene [12], and fullerene [13, 14]. In these studies, the JT vibrational modes that strongly couple to the degenerate electronic states are identified. Various phenomena are related to vibronic coupling other than the JT effect. The applicability of VCD has been extended to non-JT molecules of ammonia, cycloparaphenylene, as shown in Sects. 3.3 and 3.4, and naphthalene [15].

A molecular Hamiltonian $\hat{H}$ can be written as

$$\hat{H} := \hat{T}_n + \hat{T}_e + \hat{U}_{ne} + \hat{U}_{ee} + \hat{U}_{nn}, \tag{5.101}$$

where $\hat{T}_n$ means the sum of nuclear kinetic energy operators, $\hat{T}_e$ denotes the sum of electronic kinetic energy operators, $\hat{U}_{ne}$ is the sum of nuclear–electronic attractive potentials, $\hat{U}_{ee}$ stands for the sum of electronic–electronic repulsive potentials, and $\hat{U}_{nn}$ represents the sum of nuclear–nuclear repulsive potentials. In more detail, these terms are defined as

$$\hat{T}_n := \sum_A -\frac{\hbar^2}{2m_A}\left[\frac{\partial^2}{\partial X_A^2} + \frac{\partial^2}{\partial Y_A^2} + \frac{\partial^2}{\partial Z_A^2}\right], \tag{5.102}$$

$$\hat{T}_e := \sum_i -\frac{\hbar^2}{2m_e}\left[\frac{\partial^2}{\partial x_i^2} + \frac{\partial^2}{\partial y_i^2} + \frac{\partial^2}{\partial z_i^2}\right], \tag{5.103}$$

$$\hat{U}_{ne} := \sum_A \sum_i -\frac{z_A e^2}{4\pi\epsilon_0 |\boldsymbol{r}_i - \boldsymbol{R}_A|}, \tag{5.104}$$

$$\hat{U}_{ee} := \sum_{i \neq j} \frac{e^2}{4\pi\epsilon_0 |\boldsymbol{r}_i - \boldsymbol{r}_j|}, \tag{5.105}$$

$$\hat{U}_{nn} := \sum_{A \neq B} \frac{z_A z_B e^2}{4\pi\epsilon_0 |\boldsymbol{R}_A - \boldsymbol{R}_B|}, \tag{5.106}$$

where m_A denotes the mass of an Ath nucleus; $\hbar$ represents the reduced Planck's constant (Dirac constant); $\boldsymbol{R}_A$ is the position of an Ath nucleus, whose ξ-component ($\xi = x, y, z$) is denoted by X_A, Y_A, and Z_A; m_e stands for the mass of electron; $\boldsymbol{r}_i$ means the position of an ith electron, whose ξ-component ($\xi = x, y, z$) is denoted by x_i, y_i, z_i; ϵ_0 is the vacuum permittivity; z_A represents the charge of an Ath nucleus; e is the elementary charge. An electronic Hamiltonian $\hat{H}_e$ including nuclear–nuclear repulsive potentials is given by

$$\hat{H}_e := \hat{T}_e + \hat{U}_{ne} + \hat{U}_{ee} + \hat{U}_{nn}, \tag{5.107}$$

and thus $\hat{H}$ is rewritten as

$$\hat{H} = \hat{T}_n + \hat{H}_e. \tag{5.108}$$

We here define $\Psi_m(\boldsymbol{r}, \boldsymbol{R})$ as an mth eigenfunction of $\hat{H}_e$ and $E_m(\boldsymbol{R})$ as its eigenvalue,

$$\hat{H}_e \Psi_m(\boldsymbol{r}, \boldsymbol{R}) = E_m(\boldsymbol{R}) \Psi_m(\boldsymbol{r}, \boldsymbol{R}), \tag{5.109}$$

where $\boldsymbol{r}$ denotes a set of electronic positions, and $\boldsymbol{R}$ stands for a set of nuclear positions. Here, spin coordinates of electrons are abbreviated for simplicity. $\Psi_m(\boldsymbol{r}, \boldsymbol{R})$ and $E_m(\boldsymbol{R})$ are called an mth adiabatic electronic wave function and adiabatic potential energy surface, respectively. A set of $\Psi_m(\boldsymbol{r}, \boldsymbol{R})$ $(m = 0, 1, \ldots)$ is sometimes called the Born–Oppenheimer basis.

Vibronic coupling is generally defined using the crude adiabatic basis $\Psi_m(\boldsymbol{r}, \boldsymbol{R}^0)$ $(m = 0, 1, \ldots)$, whose nuclear configuration is fixed, instead of the Born–Oppenheimer one $\Psi_m(\boldsymbol{r}, \boldsymbol{R})$ $(m = 0, 1, \ldots)$. We can expand $\hat{H}$ around reference geometry $\boldsymbol{R}^0$ in the powers of mass-weighted normal coordinates Q_α $(\alpha = 1, 2, \ldots)$ as

$$\hat{H} = \hat{H}_0 + \sum_\alpha \left(\frac{\partial \hat{H}}{\partial Q_\alpha} \right)_{\boldsymbol{R}^0} Q_\alpha + \frac{1}{2} \sum_{\alpha,\beta} \left(\frac{\partial^2 \hat{H}}{\partial Q_\alpha \partial Q_\beta} \right)_{\boldsymbol{R}^0} Q_\alpha Q_\beta + \cdots, \tag{5.110}$$

where $\hat{H}_0$ is $\hat{H}$ at geometry $\boldsymbol{R}^0$, which corresponds to $Q_\alpha = 0$ $(\alpha = 1, 2, \ldots)$. This is the so-called Herzberg–Teller expansion. For an eigenfunction $\Phi(\boldsymbol{r}, \boldsymbol{R})$ of $\hat{H}$, we employ the crude adiabatic representation:

$$\Phi(\boldsymbol{r}, \boldsymbol{R}) = \sum_m \chi_m(\boldsymbol{R}) \Psi_m(\boldsymbol{r}, \boldsymbol{R}^0). \tag{5.111}$$

If we ignore the translations and rotations of a molecule, then $\Phi(\boldsymbol{r}, \boldsymbol{R})$ is a vibronic wave function. The Hamiltonian matrix $\hat{\mathbf{H}}$ of $\hat{H}$ represented by the crude adiabatic basis $\Psi_m(\boldsymbol{r}, \boldsymbol{R}^0)$ $(m = 0, 1, \ldots)$ is given by

$$\hat{\mathbf{H}} = \begin{pmatrix} \langle \Psi_0(\boldsymbol{r}, \boldsymbol{R}^0) | \hat{H} | \Psi_0(\boldsymbol{r}, \boldsymbol{R}^0) \rangle & \langle \Psi_0(\boldsymbol{r}, \boldsymbol{R}^0) | \hat{H} | \Psi_1(\boldsymbol{r}, \boldsymbol{R}^0) \rangle & \cdots \\ \langle \Psi_1(\boldsymbol{r}, \boldsymbol{R}^0) | \hat{H} | \Psi_0(\boldsymbol{r}, \boldsymbol{R}^0) \rangle & \langle \Psi_1(\boldsymbol{r}, \boldsymbol{R}^0) | \hat{H} | \Psi_1(\boldsymbol{r}, \boldsymbol{R}^0) \rangle & \cdots \\ \vdots & \vdots & \ddots \end{pmatrix}. \tag{5.112}$$

The matrix element $(\hat{\mathbf{H}})_{mn}$ is written as

$$(\hat{\mathbf{H}})_{mn} = \hat{T}_n \delta_{mn} + E_m(\boldsymbol{R}^0) \delta_{mn} + \sum_\alpha V_{mn,\alpha} Q_\alpha + \frac{1}{2} \sum_{\alpha,\beta} W_{mn,\alpha\beta} Q_\alpha Q_\beta + \cdots, \tag{5.113}$$

where a linear vibronic coupling constant (VCC), $V_{mn,\alpha}$, and quadratic VCC (QVCC), $W_{mn,\alpha\beta}$, between mth and nth states are introduced as

$$V_{mn,\alpha} := \langle \Psi_m(\boldsymbol{r}, \boldsymbol{R}^0)| \left(\frac{\partial \hat{H}}{\partial Q_\alpha} \right)_{\boldsymbol{R}^0} |\Psi_n(\boldsymbol{r}, \boldsymbol{R}^0)\rangle, \tag{5.114}$$

$$W_{mn,\alpha\beta} := \langle \Psi_m(\boldsymbol{r}, \boldsymbol{R}^0)| \left(\frac{\partial^2 \hat{H}}{\partial Q_\alpha \partial Q_\beta} \right)_{\boldsymbol{R}^0} |\Psi_n(\boldsymbol{r}, \boldsymbol{R}^0)\rangle. \tag{5.115}$$

Such a Hamiltonian matrix is called vibronic Hamiltonian. For simplicity, we hereafter represent diagonal elements of VCCs, $V_{mm,\alpha}$ and $W_{mm,\alpha\beta}$, as $V_{m,\alpha}$ and $W_{m,\alpha\beta}$, respectively.

We can obtain the adiabatic potential energy surface $E_m(\boldsymbol{R})$ from the diagonalization of the potential part after the second term in Eqs. 5.112 and 5.113. With the help of the perturbation theory, $E_m(\boldsymbol{R})$ is given by

$$E_m(\boldsymbol{R}) = E_m(\boldsymbol{R}^0) + \sum_\alpha V_{m,\alpha} Q_\alpha + \frac{1}{2} \sum_{\alpha,\beta} \left[W^m_{\alpha\beta} - 2 \sum_{n \neq m} \frac{V_{mn,\alpha} V_{nm,\beta}}{E_n(\boldsymbol{R}^0) - E_m(\boldsymbol{R}^0)} \right] Q_\alpha Q_\beta + \cdots . \tag{5.116}$$

In particular, if only the single vibrational mode α is taken into account, then $E_m(\boldsymbol{R})$ is written as

$$E_m(\boldsymbol{R}) = E_m(\boldsymbol{R}^0) + V_{m,\alpha} Q_\alpha + \frac{1}{2} \left[W_{m,\alpha\alpha} - 2 \sum_{n \neq m} \frac{V^2_{mn,\alpha}}{E_n(\boldsymbol{R}^0) - E_m(\boldsymbol{R}^0)} \right] Q^2_\alpha + \cdots . \tag{5.117}$$

Within the orbital approximation, the orbital mixing (pseudo-Jahn–Teller) term is given by

$$-\sum_{n \neq m} \frac{V_{mn,\alpha} V_{mn,\beta}}{E_n(\boldsymbol{R}^0) - E_m(\boldsymbol{R}^0)} Q_\alpha Q_\beta = - \sum_{a \in \mathrm{occ}} \sum_{b \in \mathrm{vir}} \frac{V_{ab,\alpha} V_{ab,\beta}}{\epsilon_b - \epsilon_a} Q_\alpha Q_\beta, \tag{5.118}$$

where ϵ_a denotes the ath-orbital energy at $\boldsymbol{R}^0$ and $V_{ab,\alpha}$ stands for the orbital vibronic coupling constant between ath and bth molecular orbitals for mode α defined later. Similar to $E_m(\boldsymbol{R})$, the adiabatic electronic wave function $\Psi_m(\boldsymbol{r}, \boldsymbol{R})$ is given by

$$\Psi_m(\boldsymbol{r}, \boldsymbol{R}) = \Psi_m(\boldsymbol{r}, \boldsymbol{R}^0) - \sum_\alpha \sum_{n \neq m} \Psi_n(\boldsymbol{r}, \boldsymbol{R}^0) \frac{V_{nm,\alpha}}{E_n(\boldsymbol{R}^0) - E_m(\boldsymbol{R}^0)} Q_\alpha + \cdots . \tag{5.119}$$

As described above, an electronic wave function and a potential energy surface change as a function of vibrational coordinates, Q_α ($\alpha = 1, 2, \ldots$), owing to the mixing of electronic states. This is the concept of vibronic coupling, which is the interaction between the motion of electrons and that of nuclei. The strength of vibronic coupling is estimated by $V_{mn,\alpha}$, $W_{mn,\alpha\beta}$, and higher-order ones.

$V_{m,\alpha}$, a diagonal element of vibronic coupling, is the origin of structural/vibrational relaxation [5, 6]. The following relation can be obtained based on the Hellmann–Feynman theorem [16, 17]:

$$V_{m,\alpha} = \left(\frac{\partial E_m(\boldsymbol{R})}{\partial Q_\alpha} \right)_{\boldsymbol{R}^0}. \tag{5.120}$$

Equation 5.120 shows that $V_{m,\alpha}$ is a force acting on nuclei along mode α in state $\Psi_m(\boldsymbol{r}, \boldsymbol{R}^0)$. If the Hessian is diagonalized with the vibrational analysis as

$$W_{m,\alpha\beta} - 2 \sum_{n \neq m} \frac{V_{mn,\alpha} V_{nm,\beta}}{E_n(\boldsymbol{R}^0) - E_m(\boldsymbol{R}^0)} = \begin{cases} \omega_{m,\alpha}^2 & (\alpha = \beta), \\ 0 & (\alpha \neq \beta), \end{cases} \tag{5.121}$$

then $E_m(\boldsymbol{R})$ is given by

$$\begin{aligned} E_m(\boldsymbol{R}) &\approx E_m(\boldsymbol{R}^0) + \sum_\alpha V_{m,\alpha} Q_\alpha + \frac{1}{2} \sum_\alpha \omega_{m,\alpha}^2 Q_\alpha^2 \\ &= E_m(\boldsymbol{R}^0) + \sum_\alpha \left[\frac{1}{2} \omega_{m,\alpha}^2 \left(Q_\alpha - \frac{|V_{m,\alpha}|}{\omega_{m,\alpha}^2} \right)^2 - \frac{V_{m,\alpha}^2}{2\omega_{m,\alpha}^2} \right], \end{aligned} \tag{5.122}$$

where we choose the direction of a vibrational vector such that the system is stabilized for a positive Q_α. The choice of the vibrational vectors results in negative VCCs, i.e., $V_\alpha = -|V_\alpha|$. If the value of $V_{m,\alpha}$ is not zero, molecular deformation spontaneously occurs with the displacement of $Q_\alpha = |V_{m,\alpha}|/\omega_{m,\alpha}^2$ to yield the stabilization/reorganization energy $\Delta E_{m,\alpha}^{\mathrm{stab}}$ [5, 6],

$$\Delta E_{m,\alpha}^{\mathrm{stab}} := \frac{V_{m,\alpha}^2}{2\omega_{m,\alpha}^2}. \tag{5.123}$$

The summation of $\Delta E_{m,\alpha}^{\mathrm{stab}}$ over all the normal modes gives the total stabilization energy $\Delta E_m^{\mathrm{stab}}$, defined as follows:

$$\Delta E_m^{\mathrm{stab}} := \sum_\alpha \Delta E_{m,\alpha}^{\mathrm{stab}}. \tag{5.124}$$

As described above, diagonal vibronic couplings give rise to structural relaxation in an electronic state. Figure 5.2 schematically illustrates the structural relaxation.

In contrast to diagonal vibronic couplings, off-diagonal ones give rise to internal conversion, i.e., non-radiative transition, between different electronic states [6, 18]. Within the crude adiabatic approximation, an eigenfunction Φ_{mv} of the molecular Hamiltonian $\hat{H}$ is written as follows:

$$\Phi_{mv}(\boldsymbol{r}, \boldsymbol{R}) = \chi_{mv}(\boldsymbol{R}) \Psi_m(\boldsymbol{r}, \boldsymbol{R}^0), \tag{5.125}$$

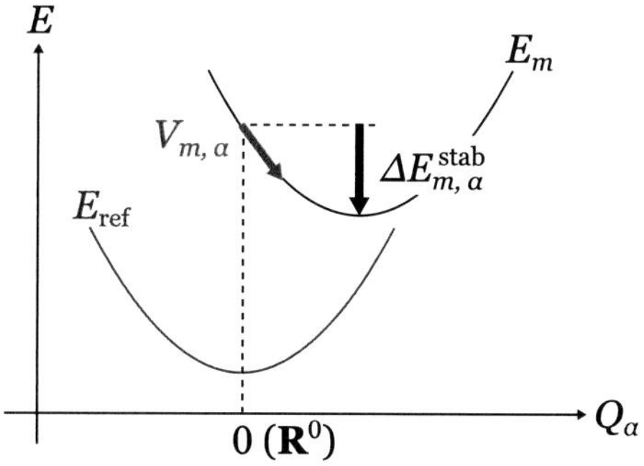

Fig. 5.2 A conceptual presentation of structural relaxation in an mth electronic state

where $\boldsymbol{R}^0$ is the equilibrium geometry in the state $\Psi_m(\boldsymbol{r}, \boldsymbol{R})$. If we ignore the translations and rotations, then $\chi_{m\nu}(\boldsymbol{R})$ is the νth vibrational wave function, and $\Phi_{m\nu}(\boldsymbol{r}, \boldsymbol{R})$ is a vibronic wave function. According to Fermi's golden rule, which describes a transition rate between different quantum states in a general manner, when perturbation $\hat{H}'$ is applied, the transition rate $k_{m\nu,n\nu'}$ from $\Phi_{m\nu}(\boldsymbol{r}, \boldsymbol{R})$ to $\Phi_{n\nu'}(\boldsymbol{r}, \boldsymbol{R})$ is given by

$$k_{m\nu,n\nu'} = \frac{2\pi}{\hbar}\rho(E)\left|\langle\Phi_{m\nu}(\boldsymbol{r}, \boldsymbol{R})|\hat{H}'|\Phi_{n\nu'}(\boldsymbol{r}, \boldsymbol{R})\rangle\right|^2, \tag{5.126}$$

where E is $E_{m\nu} - \hbar\omega$ for the radiative transition, in which $E_{m\nu}$ denotes the eigenvalue of $\hat{H}$ corresponding to $\Phi_{m\nu}(\boldsymbol{r}, \boldsymbol{R})$ and ω is the frequency of the light, and E is $E_{m\nu}$ for the internal conversion. $\rho(E)$ stands for the density of states in the final state $\Phi_{n\nu'}(\boldsymbol{r}, \boldsymbol{R})$. Therefore, the internal-conversion rate $k_{m\to n}^{\text{IC}}(T)$ from an mth electronic state to an nth electronic state at the absolute temperature T is given by

$$\begin{aligned} k_{m\to n}^{\text{IC}}(T) &= \sum_{\nu,\nu'} P_{m\nu}(T) k_{m\nu,n\nu'} \\ &= \frac{2\pi}{\hbar}\sum_{\nu,\nu'} P_{m\nu}(T)\rho(E_{m\nu}) \\ &\quad \times \left|\langle\Phi_{m\nu}(\boldsymbol{r}, \boldsymbol{R})|\hat{H}'|\Phi_{n\nu'}(\boldsymbol{r}, \boldsymbol{R})\rangle\right|^2, \end{aligned} \tag{5.127}$$

where $P_{m\nu}(T)$ denotes a statistical weight of the initial state $\Phi_{m\nu}(\boldsymbol{r}, \boldsymbol{R})$ owing to Boltzmann's distribution and $\hat{H}'$ stands for a perturbation caused by molecular vibrations:

$$\hat{H}' = \sum_\alpha \left(\frac{\partial \hat{H}}{\partial Q_\alpha}\right)_{\boldsymbol{R}^0} Q_\alpha + \cdots . \tag{5.128}$$

Within the first-order approximation, a matrix element of this perturbation is given by

$$
\begin{aligned}
\langle \Phi_{mv}(\boldsymbol{r},\boldsymbol{R})|\hat{H}'|\Phi_{nv'}(\boldsymbol{r},\boldsymbol{R})\rangle &= \sum_{\alpha} \langle \Psi_m(\boldsymbol{r},\boldsymbol{R}^0)| \left(\frac{\partial \hat{H}}{\partial Q_\alpha}\right)_{\boldsymbol{R}^0} |\Psi_n(\boldsymbol{r},\boldsymbol{R}^0)\rangle \\
&\quad \times \langle \chi_{mv}(\boldsymbol{R})|Q_\alpha|\chi_{nv'}(\boldsymbol{R})\rangle \\
&= \sum_{\alpha} V_{mn,\alpha} \langle \chi_{mv}(\boldsymbol{R})|Q_\alpha|\chi_{nv'}(\boldsymbol{R})\rangle .
\end{aligned} \tag{5.129}
$$

The substitution of Eq. 5.129 into Eq. 5.127 leads to [17]

$$
\begin{aligned}
k^{\mathrm{IC}}_{m\to n}(T) &= \frac{2\pi}{\hbar} \sum_{\alpha,\beta} V_{mn,\alpha} V_{nm,\beta} \sum_{v,v'} P_{mv}(T)\rho(E_{mv}) \\
&\quad \times \langle \chi_{mv}(\boldsymbol{R})|Q_\alpha|\chi_{nv'}(\boldsymbol{R})\rangle \langle \chi_{nv'}(\boldsymbol{R})|Q_\beta|\chi_{mv}(\boldsymbol{R})\rangle .
\end{aligned} \tag{5.130}
$$

In particular, if only the single vibrational mode α is taken into consideration, $k^{\mathrm{IC}}_{m\to n}(T)$ is given by

$$
k^{\mathrm{IC}}_{m\to n}(T) = \frac{2\pi}{\hbar} V^2_{mn,\alpha} \sum_{v,v'} P_{mv}(T)\rho(E_{mv}) \langle \chi_{mv}(\boldsymbol{R})|Q_\alpha|\chi_{nv'}(\boldsymbol{R})\rangle^2 . \tag{5.131}
$$

Equations 5.130 and 5.131 show that off-diagonal vibronic couplings, $V_{mn,\alpha}$ ($\alpha = 1, 2, \ldots$), govern the strength of internal conversion. Figure 5.3 schematically illustrates internal conversion.

We can introduce a density form of a VCC, vibronic coupling density (VCD) [5, 6, 10, 15]. It is a useful concept to analyze the origin of vibronic coupling. VCD is formulated as follows.

First, we employ the equilibrium geometry in a reference electronic state $\Psi_{\mathrm{ref}}(\boldsymbol{r},\boldsymbol{R}^0)$ as reference geometry $\boldsymbol{R}^0$. The reference geometry is optimized in the state $\Psi_{\mathrm{ref}}(\boldsymbol{r},\boldsymbol{R})$:

$$
\left(\frac{\partial E_{\mathrm{ref}}(\boldsymbol{R})}{\partial Q_\alpha}\right)_{\boldsymbol{R}^0} = \langle \Psi_{\mathrm{ref}}(\boldsymbol{r},\boldsymbol{R}^0)| \left(\frac{\partial \hat{H}}{\partial Q_\alpha}\right)_{\boldsymbol{R}^0} |\Psi_{\mathrm{ref}}(\boldsymbol{r},\boldsymbol{R}^0)\rangle = 0. \tag{5.132}
$$

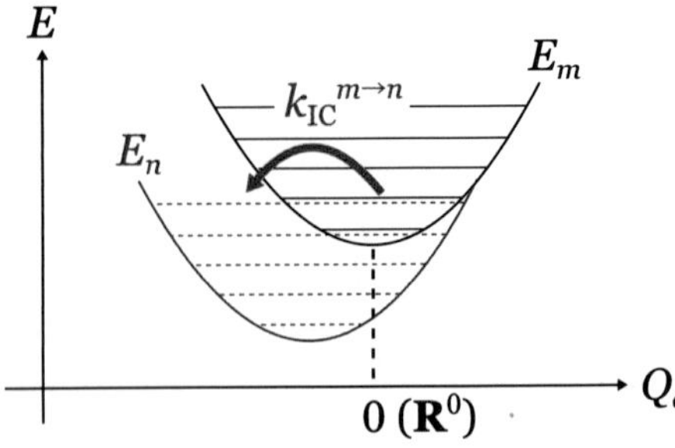

Fig. 5.3 A conceptual presentation of internal conversion from an mth electronic state to an nth electronic state

The electronic part of a linear vibronic coupling operator, $\left(\partial \hat{H}/\partial Q_\alpha\right)_{\boldsymbol{R}^0}$, can be written as follows:

$$\left(\frac{\partial \hat{H}}{\partial Q_\alpha}\right)_{\boldsymbol{R}^0} = \left(\frac{\partial \hat{U}_{ne}}{\partial Q_\alpha}\right)_{\boldsymbol{R}^0} + \left(\frac{\partial \hat{U}_{nn}}{\partial Q_\alpha}\right)_{\boldsymbol{R}^0}. \tag{5.133}$$

On the other hand, $\hat{U}_{ne}$ can be represented as the sum of one-electron potentials $u(\boldsymbol{r}_i)$ $(i = 1, 2, \ldots)$,

$$\hat{U}_{ne} = \sum_i u(\boldsymbol{r}_i), \tag{5.134}$$

where $u(\boldsymbol{r}_i)$ is defined as

$$u(\boldsymbol{r}_i) = \sum_A -\frac{Z_A e^2}{4\pi\epsilon_0 |\boldsymbol{r}_i - \boldsymbol{R}_A|}. \tag{5.135}$$

Accordingly, using a one-electron potential derivative $v_\alpha(\boldsymbol{r}_i)$,

$$v_\alpha(\boldsymbol{r}_i) := \left(\frac{\partial u(\boldsymbol{r}_i)}{\partial Q_\alpha}\right)_{\boldsymbol{R}^0}, \tag{5.136}$$

$\left(\partial \hat{H}/\partial Q_\alpha\right)_{\boldsymbol{R}^0}$ can be rewritten by

$$\left(\frac{\partial \hat{H}}{\partial Q_\alpha}\right)_{\boldsymbol{R}^0} = \sum_i v_\alpha(\boldsymbol{r}_i) + \left(\frac{\partial \hat{U}_{nn}}{\partial Q_\alpha}\right)_{\boldsymbol{R}^0}. \tag{5.137}$$

The substitution of Eq. 5.137 into Eq. 5.114 leads to

$$\begin{aligned}
V_{mn,\alpha} &= \sum_i \int \left[\int \Psi_m^*(\boldsymbol{r}, \boldsymbol{R}^0)\Psi_n(\boldsymbol{r}, \boldsymbol{R}^0) d\boldsymbol{r}_1 d\omega_1 \cdots d\boldsymbol{r}_{i-1} d\omega_{i-1} d\omega_i d\boldsymbol{r}_{i+1} d\omega_{i+1} \cdots d\boldsymbol{r}_N d\omega_N\right] \\
&\quad v_\alpha(\boldsymbol{r}_i) d\boldsymbol{r}_i + \left(\frac{\partial \hat{U}_{nn}}{\partial Q_\alpha}\right)_{\boldsymbol{R}^0} \delta_{mn} \\
&= \sum_i \int \frac{1}{N} \rho_{mn}(\boldsymbol{r}_i) v_\alpha(\boldsymbol{r}_i) d\boldsymbol{r}_i + \left(\frac{\partial \hat{U}_{nn}}{\partial Q_\alpha}\right)_{\boldsymbol{R}^0} \delta_{mn} \\
&= \int \rho_{mn}(\boldsymbol{x}) v_\alpha(\boldsymbol{x}) d\boldsymbol{x} + \left(\frac{\partial \hat{U}_{nn}}{\partial Q_\alpha}\right)_{\boldsymbol{R}^0} \delta_{mn},
\end{aligned} \tag{5.138}$$

where N is the number of electrons, and $\rho_{mn}(\boldsymbol{r}_i)$ stands for an overlap density between the mth and nth states,

$$\rho_{mn}(\boldsymbol{r}_i) := N \int \Psi_m^*(\boldsymbol{r}, \boldsymbol{R}^0)\Psi_n(\boldsymbol{r}, \boldsymbol{R}^0) d\boldsymbol{r}_1 d\omega_1 \cdots d\boldsymbol{r}_{i-1} d\omega_{i-1} d\omega_i d\boldsymbol{r}_{i+1} d\omega_{i+1} \cdots d\boldsymbol{r}_N d\omega_N, \tag{5.139}$$

in which $\boldsymbol{r}_i$ and ω_i denote the spatial and spin coordinates of an ith electron, respectively. A diagonal element of an overlap density, $\rho_{mm}(\boldsymbol{x})$, equals the total electron density in the state $\Psi_m(\boldsymbol{r}, \boldsymbol{R}^0)$. We hereafter abbreviate a diagonal element $\rho_{mm}(\boldsymbol{x})$ as $\rho_m(\boldsymbol{x})$. An off-diagonal element of $\rho_{mn}(\boldsymbol{x})$ $(m \neq n)$ is equivalent to a transition density within the orbital approximation, or the independent-particle approximation. Equations 5.132 and 5.138 are combined to yield

$$0 = \int \rho_{\text{ref}}(\boldsymbol{x}) v_\alpha(\boldsymbol{x}) d\boldsymbol{x} + \left(\frac{\partial \hat{U}_{nn}}{\partial Q_\alpha}\right)_{\boldsymbol{R}^0}, \tag{5.140}$$

where $\rho_{\text{ref}}(\boldsymbol{x})$ is the total electron density in the state $\Psi_{\text{ref}}(\boldsymbol{r}, \boldsymbol{R}^0)$. Then, subtracting Eq. 5.140 from Eq. 5.138 leads to the simplification of a diagonal linear VCC,

$$V_{m,\alpha} = \int \Delta\rho_m(\boldsymbol{x}) v_\alpha(\boldsymbol{x}) d\boldsymbol{x}, \tag{5.141}$$

where an electron-density difference $\Delta\rho_m(\boldsymbol{x})$ is defined as

$$\Delta\rho_m(\boldsymbol{x}) := \rho_m(\boldsymbol{x}) - \rho_{\text{ref}}(\boldsymbol{x}). \tag{5.142}$$

As shown in Eq. 5.141, the introduction of $\Delta\rho_m(\boldsymbol{x})$ apparently vanishes the second term, $\left(\partial \hat{U}_{nn}/\partial Q_\alpha\right)_{\boldsymbol{R}^0}$.

The obtained relations are summarized as follows. A linear VCC, $V_{mn,\alpha}$, is described as an integral of a VCD, $\eta_{mn,\alpha}(\boldsymbol{x})$ [5, 6, 10, 15],

$$V_{mn,\alpha} = \int \eta_{mn,\alpha}(\boldsymbol{x}) d\boldsymbol{x}, \tag{5.143}$$

$$\eta_{mn,\alpha}(\boldsymbol{x}) := \begin{cases} \Delta\rho_m(\boldsymbol{x}) \times v_\alpha(\boldsymbol{x}) \ (m = n), \\ \rho_{mn}(\boldsymbol{x}) \times v_\alpha(\boldsymbol{x}) \ \ (m \neq n). \end{cases} \tag{5.144}$$

The concept of VCD enables us to visually and intuitively analyze the origin of vibronic coupling in terms of $\Delta\rho_m(\boldsymbol{x})/\rho_{mn}(\boldsymbol{x})$, which is obtained from electronic structures, and $v_\alpha(\boldsymbol{x})$, which is obtained from vibrational structures.

5.4.2 Higher-Order Vibronic Coupling Density

Similar to a linear VCC, QVCC and higher-order VCCs can also be represented in density forms [6]. A kth-order VCC $W_{mn,\alpha_1\cdots\alpha_k}$ is defined by

$$W_{mn,\alpha_1\cdots\alpha_k} := \langle\Psi_m(\boldsymbol{r},\boldsymbol{R}^0)|\left(\frac{\partial^k\hat{H}}{\partial Q_{\alpha_1}\cdots\partial Q_{\alpha_k}}\right)_{\boldsymbol{R}^0}|\Psi_n(\boldsymbol{r},\boldsymbol{R}^0)\rangle. \tag{5.145}$$

In particular, $W_{mn,\alpha}$ is $V_{mn,\alpha}$. $\left(\partial^k\hat{H}/\partial Q_{\alpha_1}\cdots\partial Q_{\alpha_k}\right)_{\boldsymbol{R}^0}$ can be rewritten by

$$\left(\frac{\partial^k\hat{H}}{\partial Q_{\alpha_1}\cdots\partial Q_{\alpha_k}}\right)_{\boldsymbol{R}^0} = \sum_i w_{\alpha_1\cdots\alpha_k}(\boldsymbol{r}_i) + \left(\frac{\partial^k\hat{U}_{nn}}{\partial Q_{\alpha_1}\cdots\partial Q_{\alpha_k}}\right)_{\boldsymbol{R}^0}, \tag{5.146}$$

where $w_{\alpha_1\cdots\alpha_k}(\boldsymbol{r}_i)$ is a kth one-electron potential derivative defined by

$$w_{\alpha_1\cdots\alpha_k}(\boldsymbol{r}_i) := \left(\frac{\partial^k u(\boldsymbol{r}_i)}{\partial Q_{\alpha_1}\cdots\partial Q_{\alpha_k}}\right)_{\boldsymbol{R}^0}. \tag{5.147}$$

The substitution of Eq. 5.146 into Eq. 5.145 leads to

$$\begin{aligned}
W_{mn,\alpha_1\cdots\alpha_k} &= \sum_i \int \left[\int \Psi_m^*(\boldsymbol{r},\boldsymbol{R}^0)\Psi_n(\boldsymbol{r},\boldsymbol{R}^0)d\boldsymbol{r}_1 d\omega_1\cdots d\boldsymbol{r}_{i-1}d\omega_{i-1}d\omega_i d\boldsymbol{r}_{i+1}d\omega_{i+1}\cdots d\boldsymbol{r}_N d\omega_N\right] \\
&\quad \times w_{\alpha_1\cdots\alpha_k}(\boldsymbol{r}_i)d\boldsymbol{r}_i + \left(\frac{\partial^k\hat{U}_{nn}}{\partial Q_{\alpha_1}\cdots\partial Q_{\alpha_k}}\right)_{\boldsymbol{R}^0}\delta_{mn} \\
&= \sum_i \int \frac{1}{N}\rho_{mn}(\boldsymbol{r}_i)w_{\alpha_1\cdots\alpha_k}(\boldsymbol{r}_i)d\boldsymbol{r}_i + \left(\frac{\partial^k\hat{U}_{nn}}{\partial Q_{\alpha_1}\cdots\partial Q_{\alpha_k}}\right)_{\boldsymbol{R}^0}\delta_{mn} \\
&= \int \rho_{mn}(\boldsymbol{x})w_{\alpha_1\cdots\alpha_k}(\boldsymbol{x})d\boldsymbol{x} + \left(\frac{\partial^k\hat{U}_{nn}}{\partial Q_{\alpha_1}\cdots\partial Q_{\alpha_k}}\right)_{\boldsymbol{R}^0}\delta_{mn}.
\end{aligned} \tag{5.148}$$

Therefore, $W_{mn,\alpha_1\cdots\alpha_k}$ is an integral of a kth-order VCD $\eta_{mn,\alpha_1\cdots\alpha_k}(\boldsymbol{x})$:

$$W_{mn,\alpha_1\cdots\alpha_k} = \int \eta_{mn,\alpha_1\cdots\alpha_k}(\boldsymbol{x})d\boldsymbol{x} + \left(\frac{\partial^k\hat{U}_{nn}}{\partial Q_{\alpha_1}\cdots\partial Q_{\alpha_k}}\right)_{\boldsymbol{R}^0}\delta_{mn}, \tag{5.149}$$

where

$$\eta_{mn,\alpha_1\cdots\alpha_k}(\boldsymbol{x}) := \rho_{mn}(\boldsymbol{x}) \times w_{\alpha_1\cdots\alpha_k}(\boldsymbol{x}). \tag{5.150}$$

5.4.3 One-Electron Property Density

We herein review vibronic coupling density from a general standpoint, leading to the concept of one-electron property density. For an N-electron state $\Psi_m(\boldsymbol{r}_1\omega_1, \boldsymbol{r}_2\omega_2, \ldots, \boldsymbol{r}_N\omega_N)$, the anti-symmetric principle for Fermions states that

$$\begin{aligned}
&\Psi_m(\boldsymbol{r}_1\omega_{\sigma(1)}, \boldsymbol{r}_2\omega_{\sigma(2)}, \ldots, \boldsymbol{x}\omega_{\sigma(i)=1}, \ldots, \boldsymbol{r}_N\omega_{\sigma(N)}) \\
&= \mathrm{sgn}(\sigma)\,\Psi_m(\boldsymbol{x}\omega_1, \boldsymbol{r}_2\omega_2, \ldots, \boldsymbol{r}_N\omega_N),
\end{aligned} \tag{5.151}$$

where $\sigma \in S_N$. The product of two electronic wave functions satisfies

$$\begin{aligned}
&\Psi_m^*(\boldsymbol{r}_{\sigma(1)}, \ldots, \boldsymbol{x}\omega_1, \ldots, \boldsymbol{r}_N\omega_{\sigma(N)})\, \Psi_n(\boldsymbol{r}_{\sigma(1)} \ldots, \boldsymbol{x}\omega_1, \ldots, \boldsymbol{r}_N\omega_{\sigma(N)}) \\
&= \{\mathrm{sgn}(\sigma)\}^2\ \Psi_m^*(\boldsymbol{x}\omega_1, \boldsymbol{r}_2\omega_2, \ldots, \boldsymbol{r}_N\omega_N)\Psi_n(\boldsymbol{x}\omega_1, \boldsymbol{r}_2\omega_2, \ldots, \boldsymbol{r}_N\omega_N) \\
&= \Psi_m^*(\boldsymbol{x}\omega_1, \boldsymbol{r}_2\omega_2, \ldots, \boldsymbol{r}_N\omega_N)\Psi_n(\boldsymbol{x}\omega_1, \boldsymbol{r}_2\omega_2, \ldots, \boldsymbol{r}_N\omega_N).
\end{aligned} \tag{5.152}$$

We can therefore obtain the following relation:

$$\begin{aligned}
&\int \Psi_m^*(\boldsymbol{x}\,\omega_1, \boldsymbol{r}_2\omega_2, \ldots, \boldsymbol{r}_N\omega_N)\Psi_n(\boldsymbol{x}\,\omega_1, \boldsymbol{r}_2\omega_2, \ldots, \boldsymbol{r}_N\omega_N)d\omega_1\, d\boldsymbol{r}_2 \cdots d\boldsymbol{r}_N \\
&= \int \Psi_m^*(\boldsymbol{r}_1\omega_1, \boldsymbol{x}\,\omega_2, \ldots, \boldsymbol{r}_N\omega_N)\Psi_n(\boldsymbol{r}_1\omega_1, \boldsymbol{x}\,\omega_2, \ldots, \boldsymbol{r}_N\omega_N) \\
&\quad d\boldsymbol{r}_1 d\omega_1\, d\omega_2\, d\boldsymbol{r}_3 d\omega_3 \cdots d\boldsymbol{r}_N d\omega_N \\
&= \cdots \\
&= \int \Psi_m^*(\boldsymbol{r}_1\omega_1, \boldsymbol{r}_2\omega_2, \ldots, \boldsymbol{r}_{N-1}\omega_{N-1}, \boldsymbol{x}\,\omega_N)\Psi_n(\boldsymbol{r}_1\omega_1, \boldsymbol{r}_2\omega_2, \ldots, \boldsymbol{r}_{N-1}\omega_{N-1}, \boldsymbol{x}\,\omega_N) \\
&\quad d\boldsymbol{r}_1 d\omega_1 \cdots d\boldsymbol{r}_{N-1} d\omega_{N-1}\, d\omega_N \\
&= \frac{1}{N}\rho_{mn}(\boldsymbol{x}).
\end{aligned} \tag{5.153}$$

Suppose that an operator $\hat{O}$ consists of one-electron operators $\hat{o}$ without any differential operators,

$$\hat{O} = \sum_{i=1}^{N} \hat{o}(\boldsymbol{r}_i), \tag{5.154}$$

where $\boldsymbol{r}_i$ denotes the spatial coordinate of electron i. A matrix element of $\hat{O}$ is given by [19]

$$\begin{aligned}
O_{mn} &= \int \cdots \int \Psi_m^* \hat{O} \Psi_n d\boldsymbol{r}_1 \cdots d\boldsymbol{r}_N \\
&= \sum_{i=1}^{N} \int \left[\int \cdots \int \Psi_m{}^* \Psi_n d\boldsymbol{r}_1 \cdots d\boldsymbol{r}_{i-1} d\omega_i d\boldsymbol{r}_{i+1} \cdots d\boldsymbol{r}_N \right] \hat{o}(\boldsymbol{r}_i) d\boldsymbol{r}_i \\
&= \sum_{i=1}^{N} \int \left[\frac{1}{N} \rho_{mn}(\boldsymbol{r}_i) \right] \hat{o}(\boldsymbol{r}_i) d\boldsymbol{r}_i \\
&= \int \Omega_{mn}(\boldsymbol{x}) d\boldsymbol{x},
\end{aligned} \tag{5.155}$$

where $\boldsymbol{r}_i = (\boldsymbol{r}_i, s_i)$ with spatial coordinate $\boldsymbol{r}_i$ and spin coordinate s_i for electron i. Additionally, a one-electron property density $\Omega_{mn}(\boldsymbol{x})$ is defined as

$$\Omega_{mn}(\boldsymbol{x}) := \rho_{mn}(\boldsymbol{x}) \times \hat{o}(\boldsymbol{x}). \tag{5.156}$$

We can therefore discuss a one-electron property in the density form.

The vibronic coupling constant $V_{mn,\alpha}$ is

$$\begin{aligned} V_{mn,\alpha} &= \int \cdots \int \Psi_m^* \left(\frac{\partial \hat{U}_{ne}}{\partial Q_\alpha} \right)_{\boldsymbol{R}_0} \Psi_n d\boldsymbol{r}_1 \cdots d\boldsymbol{r}_N \\ &+ \int \cdots \int \Psi_m^* \left(\frac{\partial \hat{U}_{nn}}{\partial Q_\alpha} \right)_{\boldsymbol{R}_0} \Psi_n d\boldsymbol{r}_1 \cdots d\boldsymbol{r}_N \\ &= \int \cdots \int \Psi_m^* \left(\frac{\partial \hat{U}_{ne}}{\partial Q_\alpha} \right)_{\boldsymbol{R}_0} \Psi_n d\boldsymbol{r}_1 \cdots d\boldsymbol{r}_N + \left(\frac{\partial \hat{U}_{nn}}{\partial Q_\alpha} \right)_{\boldsymbol{R}^0} \delta_{mn}, \end{aligned} \quad (5.157)$$

where U_{ne} denotes the sum of nuclear–electronic potentials, U_{nn} stands for the sum of nuclear–nuclear potentials, and Q_α is a mass-weighted normal coordinate of mode α. Since $(\partial U_{ne}/\partial Q_\alpha)_{\boldsymbol{R}_0}$ can be written as the sum of one-electron potential derivatives:

$$\left(\frac{\partial U_{ne}}{\partial Q_\alpha} \right)_{\boldsymbol{R}_0} = \sum_{i=1}^{N} v_\alpha(\boldsymbol{r}_i), \quad (5.158)$$

$V_{mn,\alpha}$ is given by the integral of the vibronic coupling density $\eta_{mn,\alpha}(\boldsymbol{x}) = \rho_{mn}(\boldsymbol{x}) \times v_\alpha(\boldsymbol{x})$ [18, 19], except the term of $\left(\partial \hat{U}_{nn}/Q_\alpha\right)_{\boldsymbol{R}^0} \delta_{mn}$.

Let us consider an electric dipole moment operator

$$\hat{\boldsymbol{\mu}} = \sum_i e\boldsymbol{r}_i, \quad (5.159)$$

which is another example of one-electron operators: $\hat{o} = e\,\boldsymbol{r}_i$. We define an electric dipole moment density as

$$\tau_{mn}(\boldsymbol{x}) = \rho_{mn}(\boldsymbol{x})\, e\boldsymbol{x}. \quad (5.160)$$

For an electronic state Ψ_0, the diagonal element of $\hat{\boldsymbol{\mu}}$ yields a permanent electric dipole moment:

$$\boldsymbol{\mu} = \int \rho_0(\boldsymbol{x})\, e\boldsymbol{x} d\boldsymbol{x} = \int \tau_{00}(\boldsymbol{x}) d\boldsymbol{x}. \quad (5.161)$$

The integrand is called a permanent dipole moment density.

The off-diagonal element of $\hat{\boldsymbol{\mu}}$ between Ψ_m and Ψ_n is a transition dipole moment.

$$\boldsymbol{\mu}_{mn} = \int \rho_{mn}(\boldsymbol{x})\, e\boldsymbol{x} d\boldsymbol{x} = \int \tau_{mn}(\boldsymbol{x}) d\boldsymbol{x}, \quad (5.162)$$

where $\tau_{mn}(\boldsymbol{x})$ is called the transition dipole moment density (TDMD). The introduction of TDMD enables us to analyze and control optical processes via electronic structures of molecules because absorptions, emissions, and scattering are governed by transition dipole moments.

5.4.4 Related Concepts to Vibronic Coupling Density

5.4.4.1 Atomic Vibronic Coupling Constant

The decomposition of $V_{mn,\alpha}$ into its atomic components would be useful for grabbing a local picture of vibronic coupling based on the evaluation. The partial derivative $\partial/\partial Q_\alpha$ is given by

$$\begin{aligned}\frac{\partial}{\partial Q_\alpha} &= \sum_A \left[\frac{\partial X_A}{\partial Q_\alpha}\frac{\partial}{\partial X_A} + \frac{\partial Y_A}{\partial Q_\alpha}\frac{\partial}{\partial Y_A} + \frac{\partial Z_A}{\partial Q_\alpha}\frac{\partial}{\partial Z_A}\right] \\ &= \sum_A \left[\frac{u_{\alpha(Ax)}}{\sqrt{m_A}}\frac{\partial}{\partial X_A} + \frac{u_{\alpha(Ay)}}{\sqrt{m_A}}\frac{\partial}{\partial Y_A} + \frac{u_{\alpha(Az)}}{\sqrt{m_A}}\frac{\partial}{\partial Z_A}\right],\end{aligned} \tag{5.163}$$

because of the following relation:

$$\begin{aligned}X_A &= X_A^0 + \sum_\alpha \frac{u_{\alpha(Ax)}}{\sqrt{m_A}} Q_\alpha, \\ Y_A &= Y_A^0 + \sum_\alpha \frac{u_{\alpha(Ax)}}{\sqrt{m_A}} Q_\alpha, \\ Z_A &= Z_A^0 + \sum_\alpha \frac{u_{\alpha(Ax)}}{\sqrt{m_A}} Q_\alpha,\end{aligned} \tag{5.164}$$

where X_A, Y_A, and Z_A are the ξ-components ($\xi = x, y, z$) of the position of an Ath nucleus; X_A^0 is X_A at $\boldsymbol{R}^0$, and $u_{\alpha(A\xi)}$ is the component of $\boldsymbol{u}_\alpha$ for an Ath nucleus and ξ $(= x, y, z)$. Accordingly, $V_{mn,\alpha}$ is rewritten by

$$\begin{aligned}V_{mn,\alpha} = \sum_A \Bigg[&\langle \Psi_m(\boldsymbol{r}, \boldsymbol{R}^0)| \left(\frac{\partial \hat{H}}{\partial X_A}\right)_{\boldsymbol{R}^0} |\Psi_n(\boldsymbol{r}, \boldsymbol{R}^0)\rangle \frac{u_{\alpha(Ax)}}{\sqrt{m_A}} \\ &+ \langle \Psi_m(\boldsymbol{r}, \boldsymbol{R}^0)| \left(\frac{\partial \hat{H}}{\partial Y_A}\right)_{\boldsymbol{R}^0} |\Psi_n(\boldsymbol{r}, \boldsymbol{R}^0)\rangle \frac{u_{\alpha(Ay)}}{\sqrt{m_A}} \\ &+ \langle \Psi_m(\boldsymbol{r}, \boldsymbol{R}^0)| \left(\frac{\partial \hat{H}}{\partial Z_A}\right)_{\boldsymbol{R}^0} |\Psi_n(\boldsymbol{r}, \boldsymbol{R}^0)\rangle \frac{u_{\alpha(Az)}}{\sqrt{m_A}} \Bigg].\end{aligned} \tag{5.165}$$

Thus, $V_{mn,\alpha}$ can be regarded as the sum of $V_{mn,\alpha(A)}$ over all the nuclei:

$$
\begin{aligned}
V_{mn,\alpha(A)} := & \langle \Psi_m(\boldsymbol{r}, \boldsymbol{R}^0) | \left(\frac{\partial \hat{H}}{\partial X_A} \right)_{\boldsymbol{R}^0} | \Psi_n(\boldsymbol{r}, \boldsymbol{R}^0) \rangle \frac{u_{\alpha(Ax)}}{\sqrt{m_A}} \\
& + \langle \Psi_m(\boldsymbol{r}, \boldsymbol{R}^0) | \left(\frac{\partial \hat{H}}{\partial Y_A} \right)_{\boldsymbol{R}^0} | \Psi_n(\boldsymbol{r}, \boldsymbol{R}^0) \rangle \frac{u_{\alpha(Ay)}}{\sqrt{m_A}} \\
& + \langle \Psi_m(\boldsymbol{r}, \boldsymbol{R}^0) | \left(\frac{\partial \hat{H}}{\partial Z_A} \right)_{\boldsymbol{R}^0} | \Psi_n(\boldsymbol{r}, \boldsymbol{R}^0) \rangle \frac{u_{\alpha(Az)}}{\sqrt{m_A}}, \qquad (5.166)
\end{aligned}
$$

where $V_{mn,\alpha(A)}$ is called an atomic VCC (AVCC) .

Instead of AVCC, we can also use regional VCD for decomposing $V_{mn,\alpha}$ into its atomic components . It is defined as the integration of VCD over the Wigner–Seitz cell of each nucleus, which is the smallest polyhedron formed by planes that perpendicularly bisect the segments between the nucleus and the neighboring nuclei.

5.4.4.2 Orbital Vibronic Coupling Density

For a Slater determinant Φ_0 constructed by orthonormal spin orbitals, $\{\chi(\boldsymbol{r}_i, \omega_i) = \psi(\boldsymbol{r}_i)\sigma(\omega_i)\}$, a diagonal element of the one-electron property can be decomposed as [20]

$$
\langle \Phi_0 | \hat{O} | \Phi_0 \rangle = \sum_{a \in \mathrm{occ}} n_a \langle \psi_a(\boldsymbol{r}_i) | \hat{o}(\boldsymbol{r}_i) | \psi_a(\boldsymbol{r}_i) \rangle,
$$

where n_a is the occupation number of ψ_a. For an off-diagonal element between Φ_0 and Φ_a^b with orbital ψ_a in Φ_0 replaced by ψ_b,

$$
\langle \Phi_0 | \hat{O} | \Phi_a^b \rangle = \langle \psi_a(\boldsymbol{x}) | \hat{o}(\boldsymbol{x}) | \psi_b(\boldsymbol{x}) \rangle
$$

. We define orbital one-electron property density $\omega_{ab}(\boldsymbol{x})$ as

$$
\omega_{ab}(\boldsymbol{x}) := \psi_a^*(\boldsymbol{x}) \psi_b(\boldsymbol{x}) \hat{o}(\boldsymbol{x}) = \rho_{ab}(\boldsymbol{x}) \times \hat{o}(\boldsymbol{x}),
$$

where $\rho_{ab}(\boldsymbol{x})$ denotes orbital overlap density (transition density). We can decompose orbital one-electron property in the density form:

$$
\langle \psi_a(\boldsymbol{x}) | \hat{o}(\boldsymbol{x}) | \psi_b(\boldsymbol{x}) \rangle = \int \omega_{ab}(\boldsymbol{x}) d\boldsymbol{x}. \qquad (5.167)
$$

A kth-order orbital vibronic coupling constant (OVCC) between orbitals a and b can be expressed in a density form:

$$
V_{ab,\alpha_1 \cdots \alpha_k} := \langle \psi_a(\boldsymbol{x}) | w_{\alpha_1 \cdots \alpha_k}(\boldsymbol{x}) | \psi_b(\boldsymbol{x}) \rangle = \int \eta_{ab,\alpha_1 \cdots \alpha_k}(\boldsymbol{x}) d\boldsymbol{x}, \qquad (5.168)
$$

where orbital vibronic coupling density (OVCD) is defined as

$$\eta_{ab,\alpha_1\cdots\alpha_k}(\boldsymbol{x}) := \rho_{ab}(\boldsymbol{x}) \times w_{\alpha_1\cdots\alpha_k}(\boldsymbol{x}). \tag{5.169}$$

5.4.4.3 Reduced Vibronic Coupling Density

Within the linear combination of atomic orbitals (LCAO) approximation, an overlap density, ρ_{mn}, is given by a linear combination of the products of two basis functions based on atomic orbitals (AOs),

$$\rho_{mn}(\boldsymbol{x}) = \sum_{A,B}\sum_{\mu,\nu} D^{mn}_{A\mu B\nu}\chi_{A\mu}(\boldsymbol{x})\chi_{B\nu}(\boldsymbol{x}), \tag{5.170}$$

where $\chi_{A\mu}$ denotes the μth basis function belonging to nucleus A. An electron-density difference, $\Delta\rho_m$, is similar to the overlap density. Accordingly, the kth-order VCD, $\eta_{mn,\alpha_1\cdots\alpha_k}$, can be written by

$$\eta_{mn,\alpha_1\cdots\alpha_k}(\boldsymbol{x}) = \sum_{A,B}\sum_{\mu,\nu} D^{mn}_{A\mu B\nu}\chi_{A\mu}(\boldsymbol{x})\chi_{B\nu}(\boldsymbol{x})w_{\alpha_1\cdots\alpha_k}(\boldsymbol{x}), \tag{5.171}$$

where $w_{\alpha_1\cdots\alpha_k}(\boldsymbol{x})$ is the kth-order potential derivative. The kth-order VCC, $W_{mn,\alpha_1\cdots\alpha_k}$, can be obtained by its spatial integration,

$$W_{mn,\alpha_1\cdots\alpha_k} = \sum_{A,B}\sum_{\mu,\nu} D^{mn}_{A\mu B\nu}\int \chi_{A\mu}(\boldsymbol{x})\chi_{B\nu}(\boldsymbol{x})w_{\alpha_1\cdots\alpha_k}(\boldsymbol{x})d\boldsymbol{x}. \tag{5.172}$$

In general, $w_{\alpha_1\cdots\alpha_k}(\boldsymbol{x})$ has symmetric distribution near atoms. In particular, the linear one, $v_\alpha(\boldsymbol{x})$, has p-type distribution around each atom, which is locally transformed as $\xi = x, y, z$:

$$v_\alpha(\boldsymbol{x}) = \sum_A \frac{z_A e^2}{4\pi\epsilon_0}\left[\frac{x - X_A}{|\boldsymbol{x} - \boldsymbol{R}_A|^3}\frac{u_{\alpha(Ax)}}{\sqrt{m_A}} + \frac{y - Y_A}{|\boldsymbol{x} - \boldsymbol{R}_A|^3}\frac{u_{\alpha(Ay)}}{\sqrt{m_A}} + \frac{z - Z_A}{|\boldsymbol{x} - \boldsymbol{R}_A|^3}\frac{u_{\alpha(Az)}}{\sqrt{m_A}}\right], \tag{5.173}$$

where z_A denotes the charge of an Ath nucleus, X_A, Y_A, and Z_A are the ξ components ($\xi = x, y, z$) of the position of an Ath nucleus; and $u_{\alpha(A\xi)}$ is the component of the normal vector $\boldsymbol{u}_\alpha$ for an Ath nucleus and ξ $(= x, y, z)$. This gives rise to cancellation due to the symmetrical reason. For example, if both $\chi_{A\mu}(\boldsymbol{x})$ and $\chi_{B\nu}(\boldsymbol{x})$ are s-type, then the integrand, $\chi_{A\mu}(\boldsymbol{x})\chi_{B\nu}(\boldsymbol{x})v_\alpha(\boldsymbol{x})$, is p-type $(= s \times s \times p)$ whose integral is exactly zero. We can eliminate such a zero contribution $\eta_{mn,\alpha_1\cdots\alpha_k}{}'(\boldsymbol{x})$ from $\eta_{mn,\alpha_1\cdots\alpha_k}(\boldsymbol{x})$ to define the reduced vibronic coupling density (RVCD), $\bar{\eta}_{mn,\alpha_1\cdots\alpha_k}(\boldsymbol{x})$, which describes net contributions to the VCC [21]:

$$\bar{\eta}_{mn,\alpha_1\cdots\alpha_k}(\boldsymbol{x}) = \eta_{mn,\alpha_1\cdots\alpha_k}(\boldsymbol{x}) - \eta_{mn,\alpha_1\cdots\alpha_k}{}'(\boldsymbol{x}). \tag{5.174}$$

5.4.4.4 Effective Vibronic Coupling Density

According to the Hellmann–Feynman theorem [16, 17], the total differential of the total electronic energy E is written by

$$dE = \sum_{\alpha} \left(\frac{\partial E}{\partial Q_{\alpha}} \right)_{\boldsymbol{R}^0} dQ_{\alpha} = \sum_{\alpha} V_{\alpha} dQ_{\alpha}, \tag{5.175}$$

where Q_{α} is a mass-weighted normal coordinate of mode α. We here introduce gradient and displacement vectors, $\boldsymbol{V}$ and $d\boldsymbol{Q}$, as

$$\boldsymbol{V} := \sum_{\alpha} V_{\alpha} \boldsymbol{u}_{\alpha}, \quad d\boldsymbol{Q} := \sum_{\alpha} \boldsymbol{u}_{\alpha} dQ_{\alpha}, \tag{5.176}$$

where $\boldsymbol{u}_{\alpha}$ is a vibrational vector of mode α, which is orthonormal with each other. Thereby, dE can be rewritten as

$$dE = \boldsymbol{V} \cdot d\boldsymbol{Q}. \tag{5.177}$$

The Cauchy–Schwarz inequality states that $|dE|$ has the maximum value of $|\boldsymbol{V}||d\boldsymbol{Q}|$ when $\boldsymbol{V}$ and $d\boldsymbol{Q}$ are parallel. Such a displacement vector is given by

$$d\boldsymbol{Q} = \boldsymbol{u}_s ds, \quad \boldsymbol{u}_s := \frac{\boldsymbol{V}}{|\boldsymbol{V}|} = \sum_{\alpha} \frac{V_{\alpha}}{\sqrt{\sum_{\alpha'} V_{\alpha'}^2}} \boldsymbol{u}_{\alpha}. \tag{5.178}$$

This is the steepest descent direction of E. The direction vector $\boldsymbol{u}_s$ with the corresponding coordinate s is called the effective mode. Since each element of $d\boldsymbol{Q}$ is given by

$$dQ_{\alpha} = \frac{V_{\alpha}}{\sqrt{\sum_{\alpha'} V_{\alpha'}^2}} ds, \tag{5.179}$$

ds is expressed as

$$ds = \sum_{\alpha} \frac{V_{\alpha}}{\sqrt{\sum_{\alpha'} V_{\alpha'}^2}} dQ_{\alpha}. \tag{5.180}$$

The VCD for the effective mode, which is called the effective VCD (EVCD), is also introduced as

$$\eta_s(\boldsymbol{x}) := \Delta\rho(\boldsymbol{x}) \times v_s(\boldsymbol{x}), \quad v_s(\boldsymbol{x}) := \left(\frac{\partial u(\boldsymbol{x})}{\partial s} \right)_{\boldsymbol{R}^0}. \tag{5.181}$$

The spatial integration of $\eta_s(\boldsymbol{x})$ yields the VCC for the effective mode,

$$V_s := \langle \Psi(\boldsymbol{r}, \boldsymbol{R}^0) | \left(\frac{\partial \hat{H}}{\partial s} \right)_{\boldsymbol{R}^0} | \Psi(\boldsymbol{r}, \boldsymbol{R}^0) \rangle. \tag{5.182}$$

It is called the effective VCC (EVCC) .

5.4.5 Vibronic Coupling Density as a Chemical Reactivity Index

A VCD can be formulated as a reactivity index based on Parr and Yang's theory [22, 23]. Before describing the VCD theory for chemical reactions, we briefly review the conventional Parr and Yang's theory to make this document self-contained. They derived the frontier orbital theory in conceptual density functional theory (DFT) [24] without the orbital approximation.

The total electronic energy of a reactant, E, is a functional of the number of total electrons, N, and a one-electron nuclear–electronic potential, $u(\boldsymbol{x})$,

$$E = E[N, u(\boldsymbol{x})]. \tag{5.183}$$

Its differential, dE, is given by

$$\begin{aligned} dE &= \left(\frac{\partial E}{\partial N} \right)_u dN + \int \left(\frac{\delta E}{\delta u} \right)_N du(\boldsymbol{x}) d\boldsymbol{x} \\ &= \mu dN + \int \rho(\boldsymbol{x}) du(\boldsymbol{x}) d\boldsymbol{x}, \end{aligned} \tag{5.184}$$

where an electronic chemical potential, μ, is defined as

$$\mu := \left(\frac{\partial E}{\partial N} \right)_u, \tag{5.185}$$

and $\rho(\boldsymbol{x})$ denotes the total electron density, which is equal to a functional derivative of E for u as per the first-order perturbation theory,

$$\rho(\boldsymbol{x}) = \left(\frac{\delta E}{\delta u} \right)_N. \tag{5.186}$$

As the same as E, μ is a functional of N and $u(\boldsymbol{x})$,

$$\mu = \mu[N, u(\boldsymbol{x})]. \tag{5.187}$$

Its differential $d\mu$ is given by

$$
\begin{aligned}
d\mu &= \left(\frac{\partial \mu}{\partial N}\right)_u dN + \int \left(\frac{\delta \mu}{\delta u}\right)_N du(\boldsymbol{x})d\boldsymbol{x} \\
&= 2\eta dN + \int f(\boldsymbol{x})du(\boldsymbol{x})d\boldsymbol{x},
\end{aligned} \tag{5.188}
$$

where absolute hardness η is defined as

$$
\eta := \frac{1}{2}\left(\frac{\partial^2 E}{\partial N^2}\right)_u, \tag{5.189}
$$

and a Fukui function $f(\boldsymbol{x})$ is defined by

$$
f(\boldsymbol{x}) := \left(\frac{\delta \mu}{\delta u}\right)_N, \tag{5.190}
$$

which is equal to a derivative of $\rho(\boldsymbol{x})$ with respect to N,

$$
f(\boldsymbol{x}) = \left(\frac{\partial \rho}{\partial N}\right)_u, \tag{5.191}
$$

because of the Maxwell relation,

$$
\left(\frac{\delta \mu}{\delta u}\right)_N = \left[\frac{\delta}{\delta u}\left(\frac{\partial E}{\partial N}\right)_u\right]_N = \left[\frac{\partial}{\partial N}\left(\frac{\delta E}{\delta u}\right)_N\right]_u = \left(\frac{\partial \rho}{\partial N}\right)_u. \tag{5.192}
$$

Differentials dE and $d\mu$ have the physical meanings of the linear responses of an isolated reactant to the perturbations, dN and $\delta u(\boldsymbol{x})$, which describe a reaction process.

Parr and Yang proposed the following principle in analogy to thermodynamics: a chemical reaction occurs such that $|d\mu|$ is maximum in the early stage of a reaction. The first term on the right-hand side of Eq. 5.188 is a global quantity uniquely determined by the reactant species. Accordingly, the first term is independent of regioselectivity. On the other hand, the second term on the right-hand side of Eq. 5.188 is local. According to the principle, a chemical reaction occurs such that the second term is maximum. That is, $f(\boldsymbol{x})$ must have a large distribution around a reactive region.

Within the finite difference approximation, $f(\boldsymbol{x})$ is equal to the electron-density difference $\Delta\rho(\boldsymbol{x})$ caused by a charge transfer,

$$
f(\boldsymbol{x}) \approx \Delta\rho(\boldsymbol{x}). \tag{5.193}
$$

In addition, $f(\boldsymbol{x})$ is equal to a frontier orbital density within the frozen orbital approximation, namely, HOMO density $|\varphi_{\mathrm{HO}}|^2$ for electrophilic attacks or LUMO density $|\varphi_{\mathrm{LU}}|^2$ for nucleophilic attacks. In this way, Parr and Yang's principle yields the fron-

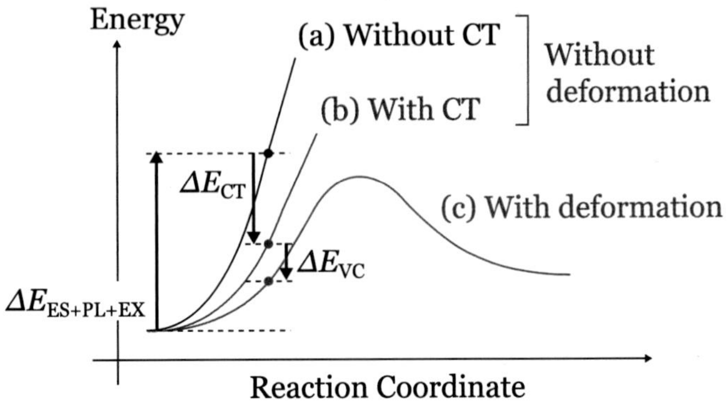

Fig. 5.4 Three types of reaction profiles. **a** $\Delta E_{ES+PL+EX}$, where only the sum of electrostatic, polarization, and exchange interactions is taken into consideration without any structural relaxation; **b** charge-transfer interaction, ΔE_{CT}, is considered along with $\Delta E_{ES+PL+EX}$ without any structural relaxation; **c** structural relaxation, ΔE_{VC}, is included in addition to all the above interactions

tier orbital theory, according to which a chemical reaction usually occurs around a region with a large HOMO/LUMO density.

We herein discuss the effect of molecular deformation based on Parr and Yang's theory (Fig. 5.4). We assume that an effective mode s coincides with the direction of the reaction path.

$$\boldsymbol{u}_s := \sum_\alpha \frac{V_\alpha}{\sqrt{\sum_{\alpha'} V_{\alpha'}^2}} \boldsymbol{u}_\alpha, \tag{5.194}$$

where $\boldsymbol{u}_\alpha$ denotes the vibrational vector of an αth normal mode. The effective mode is the steepest descent direction of the adiabatic potential surface in the charge-transfer state. Accordingly, it can be regarded as an intramolecular reaction mode. Then, $du(\boldsymbol{x})$ defined in the conceptual DFT can be expressed using a reaction coordinate s,

$$du(\boldsymbol{x}) = \left(\frac{\partial u(\boldsymbol{x})}{\partial s}\right)_{\boldsymbol{R}^0} ds + \cdots = v_s(\boldsymbol{x})ds + \cdots, \tag{5.195}$$

where $v_s(\boldsymbol{x})$ has the same definition as $v_\alpha(\boldsymbol{x})$,

$$v_s(\boldsymbol{x}) := \left(\frac{\partial u(\boldsymbol{x})}{\partial s}\right)_{\boldsymbol{R}^0}. \tag{5.196}$$

Within the finite difference approximation, Eq. 5.188 along with Eq. 5.195 can be rewritten by

$$d\mu = 2\eta dN + \int \Delta\rho(\boldsymbol{x}) v_s(\boldsymbol{x}) d\boldsymbol{x} ds. \tag{5.197}$$

The product of $\Delta\rho(\boldsymbol{x})$ and $v_s(\boldsymbol{x})$ is equal to a diagonal linear VCD $\eta_s(\boldsymbol{x})$ with respect to s,

$$\eta_s(\boldsymbol{x}) := \Delta\rho(\boldsymbol{x}) \times v_s(\boldsymbol{x}). \tag{5.198}$$

Thus, the following relation can be obtained:

$$d\mu = 2\eta dN + \int \eta_s(\boldsymbol{x}) d\boldsymbol{x} ds. \tag{5.199}$$

To predict a reactive region that gives a large $|d\mu|$, it is necessary but not sufficient for Fukui function $f(\boldsymbol{x})$ to take a large value in the reactive region. The potential derivative $v_s(\boldsymbol{x})$ also plays an important role in $|d\mu|$. Consequently, the vibronic coupling density for the reactive mode s can be a chemical reactivity index that can predict a large $|d\mu|$ region.

For example, Fig. 5.5 shows the vibronic coupling density of the naphthalene cation for the reaction mode s, defined by the steepest direction.

We can find that the vibronic coupling density has a large value near the α-carbons. This means that the motion of the α-carbon couples with the hole. This is consistent with the prediction of the frontier orbital theory.

Using the nuclear Fukui function [25, 26] defined by

$$\phi_{XA} = -\left(\frac{\partial U}{\partial X_A}\right)_N, \quad \phi_{YA} = -\left(\frac{\partial U}{\partial Y_A}\right)_N, \quad \phi_{ZA} = -\left(\frac{\partial U}{\partial Z_A}\right)_N, \tag{5.200}$$

the total differential of the chemical potential is written as

$$d\mu = 2\eta dN - \sum_A \left(\phi_{XA} dX_A + \phi_{YA} dY_A + \phi_{ZA} dZ_A\right). \tag{5.201}$$

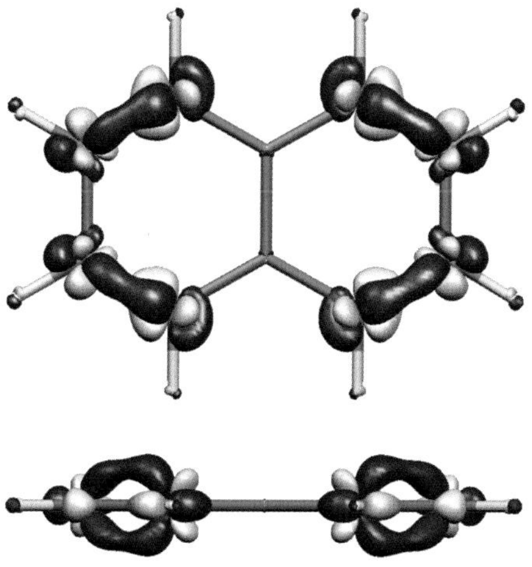

Fig. 5.5 Vibronic coupling density of the naphthalene cation. Blue and gray surfaces denote negative and positive densities, respectively. Reprinted by permission from Springer Nature, The Jahn–Teller Effect: Fundamentals and Implications for Physics and Chemistry, Springer Series in Chemical Physics, vol. 97, ed. by H. Köppel, D.R. Yarkony, H. Barentzen (Springer-Verlag, Berlin, 2009)

The mass-weighted normal coordinate is expressed in terms of the nuclear coordinate $\boldsymbol{R}$ as

$$Q_\alpha = \sum_A \mathsf{A}_{\alpha X_A} X_A + \mathsf{A}_{\alpha Y_A} Y_A + \mathsf{A}_{\alpha Z_A} Z_A. \tag{5.202}$$

Therefore,

$$d\mu = 2\eta dN + \sum_\alpha V_\alpha \sum_A \left(\mathsf{A}_{\alpha X_A} dX_A + \mathsf{A}_{\alpha Y_A} dY_A + \mathsf{A}_{\alpha Z_A} dZ_A \right), \tag{5.203}$$

and

$$\phi_{XA} = -\sum_\alpha V_\alpha \mathsf{A}_{\alpha X_A}, \quad \phi_{YA} = -\sum_\alpha V_\alpha \mathsf{A}_{\alpha Y_A}, \quad \phi_{ZA} = -\sum_\alpha V_\alpha \mathsf{A}_{\alpha Z_A}. \tag{5.204}$$

The relation between the Jahn–Teller system and the Fukui function has been discussed by Balawender et al. [26].

5.4.6 Physical Quantities Related to Vibronic Coupling Constant

In solid-state physics, the vibronic coupling is sometimes called electron–phonon or electron–lattice coupling, mainly used for crystal vibrations. Let us look at the other form of the VCC familiar with physicists. The electron–phonon coupling is defined in the second quantized form of the vibronic Hamiltonian,

$$\begin{aligned}\hat{H} &= E_0 + \sum_\alpha \left[-\frac{\hbar^2}{2} \frac{\partial^2}{\partial Q_\alpha^2} + \frac{\omega_\alpha^2}{2} Q_\alpha^2 + V_\alpha Q_\alpha \right] \\ &= E_0 \hat{c}^\dagger \hat{c} + \sum_\alpha \left[\hbar\omega_\alpha \left(\hat{b}_\alpha^\dagger \hat{b}_\alpha + \frac{1}{2} \right) + \lambda_\alpha \hat{c}^\dagger \hat{c} (\hat{b}_\alpha^\dagger + \hat{b}_\alpha) \right],\end{aligned} \tag{5.205}$$

where $\hat{c}^\dagger$ and $\hat{c}$ are the creation and annihilation operators of the one-electron state, respectively. $\hat{b}_\alpha^\dagger$ and $\hat{b}_\alpha$ are the creation and annihilation operators of the vibrational state with vibrational energy $\hbar\omega_\alpha$:

$$\hat{b}_\alpha^\dagger := \sqrt{\frac{\omega_\alpha}{2\hbar}} Q_\alpha - \frac{i}{\sqrt{2\hbar\omega_\alpha}} \left(\frac{\hbar}{i} \frac{\partial}{\partial Q_\alpha} \right), \tag{5.206}$$

$$\hat{b}_\alpha := \sqrt{\frac{\omega_\alpha}{2\hbar}} Q_\alpha + \frac{i}{\sqrt{2\hbar\omega_\alpha}} \left(\frac{\hbar}{i} \frac{\partial}{\partial Q_\alpha} \right). \tag{5.207}$$

λ_α is the electron–phonon coupling constant, the dimension of which is that of energy. Since Q_α is rewritten as

$$Q_\alpha = \sqrt{\frac{\hbar}{2\omega_\alpha}} \left(\hat{b}_\alpha^\dagger + \hat{b}_\alpha \right), \tag{5.208}$$

λ_α is given by

$$\lambda_\alpha := \sqrt{\frac{\hbar}{2\omega_\alpha}} V_\alpha. \tag{5.209}$$

The dimensionless electron–phonon coupling constant, i.e., the dimensionless VCC, is also defined by

$$g_\alpha := \frac{\lambda_\alpha}{\hbar\omega_\alpha} = \frac{V_\alpha}{\sqrt{2\hbar\omega_\alpha^3}}, \tag{5.210}$$

and the Hamiltonian becomes

$$\hat{H} = E_0 \hat{c}^\dagger \hat{c} + \hbar\omega_\alpha \left[\left(\hat{b}_\alpha^\dagger \hat{b}_\alpha + \frac{1}{2} \right) + g_\alpha \hat{c}^\dagger \hat{c} (\hat{b}_\alpha^\dagger + \hat{b}_\alpha) \right]. \tag{5.211}$$

The Rys–Huang factor is another measure of the coupling strength, which is defined by

$$S_\alpha := \frac{\Delta E_\alpha^{\text{stab}}}{\hbar\omega_\alpha} = \frac{V_\alpha^2}{2\hbar\omega_\alpha^3} = \frac{1}{2} g_\alpha^2, \tag{5.212}$$

where ΔE_α is the reorganization energy of mode α, which is equal to the Jahn–Teller stabilization energy in the case of a degenerate system. S_α is the stabilization energy measured by vibrational energy.

We also mention the relation between the VCC and mechanical force. Table 5.12 lists the fundamental physical quantities and their derivatives. Based on these quantities, physical dimensions and atomic units of VCCs are given in Table 5.13. The VCC is the derivative of energy for a mass-weighted coordinate. On the other hand, the force has the dimension of the derivative of energy for length. The kth-order partial derivatives satisfy

$$\frac{\partial^k}{\partial q_\alpha^k} = \mu_\alpha^{k/2} \frac{\partial^k}{\partial Q_\alpha^k}, \tag{5.213}$$

where q_α and Q_α stand for vibrational coordinates of mode α in the real and mass-weighted spaces, and μ_α denotes the reduced mass of the mode. Accordingly, a linear VCC, V_α, can be converted to the force, F_α, by multiplying $\sqrt{\mu_\alpha}$:

$$F_\alpha = \sqrt{\mu_\alpha} V_\alpha. \tag{5.214}$$

As with the linear VCC, higher-order VCCs can be converted to those without the dimension of mass. For example, if $\mu_\alpha = \kappa_\mu$ amu and $V_\alpha = \kappa_V$ a.u. (a.u. =

Table 5.12 Atomic units of fundamental physical quantities and their derivatives. Physical dimensions and their symbols follow the *International System of Units*

Quantity	Dimension	Atomic unit of quantity		
		Definition	Name	Expression
Mass	M	Electron mass		m_e
Charge	IT	Elementary charge		e
Action	ML^2T^{-1}	Reduced Planck's constant (Dirac constant)		$\hbar$
Length	L	Bohr radius	Bohr	a_0
energy	ML^2T^{-2}	$\frac{\hbar^2}{m_e a_0^2}$	Hartree	E_h
Time	T	$\frac{\hbar}{E_h}$		

Table 5.13 Atomic units related to vibronic coupling constants and their related quantities. Physical dimensions and their symbols follow the *International System of Units*

Quantity	Expression	Dimension	Atomic unit
Wavenumber	$\tilde{\nu}_\alpha$	L^{-1}	a_0^{-1}
Speed of light	c	LT^{-1}	$a_0\left(\frac{\hbar}{E_h}\right)^{-1}$
Frequency	$\nu_\alpha = \tilde{\nu}_\alpha c$	T^{-1}	$\left(\frac{\hbar}{E_h}\right)^{-1}$
Angular frequency	$\omega_\alpha = 2\pi\nu_\alpha$	T^{-1}	$\left(\frac{\hbar}{E_h}\right)^{-1}$
Force constant	$\kappa_\alpha = \omega_\alpha^2$	T^{-2}	$E_h m_e^{-1} a_0^{-2}$
Linear vibronic coupling constant	$V_{mn,\alpha}$	$M^{1/2}\,L\,T^{-2}$	$E_h m_e^{-1/2} a_0^{-1}$
Quadratic vibronic coupling constant	$W_{mn,\alpha\beta}$	T^{-2}	$E_h m_e^{-1} a_0^{-2}$
kth-order vibronic coupling constant	$W_{mn,\alpha_1\cdots\alpha_k}$	$M^{1-k/2}\,L^{2-k}\,T^{-2}$	$E_h m_e^{-k/2} a_0^{-k}$

$E_h m_e^{-1/2} a_0^{-1}$), then $F_\alpha = \sqrt{\kappa_\mu \kappa_V} \times 3.52 \times 10^3$ nN. In the usual case, the reduced mass is about ~ 1 to $\sim 10^2$ amu. The linear VCC of $\sim 10^{-4}$ a.u. means that the force in the order of nN is applied to nuclei in a molecule.

References

1. Ceulemans AJ (2013) Group theory applied to chemistry. Springer, Dordrecht
2. Fischer G (1984) Vibronic coupling: the interaction between the electronic and nuclear motions. Academic Press, London
3. Bersuker I, Polinger V (1989) Vibronic interactions in molecules and crystals. Springer, Berlin
4. Azumi T, Matsuzaki K (1977) Photochem Photobiol 25(3):315

5. Sato T, Tokunaga K, Iwahara N, Shizu K, Tanaka K (2009) In: Köppel H, Yarkony DR, Barentzen H (eds) The Jahn–Teller effect: fundamentals and implications for physics and chemistry, Springer series in chemical physics, vol 97. Springer, Berlin
6. Sato T, Uejima M, Iwahara N, Haruta N, Shizu K, Tanaka K (2013) J Phys: Conf Ser 428
7. Bersuker I, Gorinchoi N, Polinger V (1984) Theo Chim Acta 66:161
8. Jahn HA, Teller E (1937) Proc R Soc London A 161:220
9. Bersuker IB (2001) Chem Rev 101(4):1067
10. Sato T, Tokunaga K, Tanaka K (2006) J Chem Phys 124
11. Tokunaga K, Sato T, Tanaka K (2007) J Mol Struct 838(1–3):116
12. Tokunaga K, Sato T, Tanaka K (2006) J Chem Phys 124(15)
13. Iwahara N, Sato T, Tanaka K, Chibotaru LF (2010) Phys Rev B 82(24)
14. Iwahara N, Sato T, Tanaka K (2012) J Chem Phys 136(17)
15. Sato T, Tokunaga K, Tanaka K (2008) J Phys Chem A 112:758
16. Hellmann H (1937) Einführung in die Quantenchemie. Deuticke and Company, Leipzig
17. Feynman RP (1939) Phys Rev 56(4):340
18. Uejima M, Sato T, Yokoyama D, Tanaka K, Park JW (2014) Phys Chem Chem Phys 16(27):14244
19. Sato T, Uejima M, Iwahara N, Haruta N, Shizu K, Tanaka K (2013) J Phys: Conf Ser 428(1)
20. Szabo A, Ostlund N (1982) Modern quantum chemistry: introduction to advanced electronic structure theory. Macmillan, New York
21. Shizu K, Sato T, Tanaka K (2010) Chem Phys Lett 491:65
22. Parr RG, Yang W (1984) J Am Chem Soc 106:4049
23. Lee C, Yang W, Parr RG (1988) J Mol Struct: THEOCHEM 163:305
24. Geerlings P, Proft FD, Langenaeker W (2003) Chem Rev 103:1793
25. Cohen M (1994) J Chem Phys 101:8988
26. Balawender R, Proft F, Geerings P (2001) J Chem Phys 114:4441

Printed in the United States
by Baker & Taylor Publisher Services